Rushed
to Judgment

Power, Conflict, and Democracy:
American Politics Into the Twenty-first Century
Robert Y. Shapiro, Editor

Power, Conflict, and Democracy: American Politics Into the Twenty-first Century

Robert Y. Shapiro, Editor

This series focuses on how the will of the people and the public interest are promoted, encouraged, or thwarted. It aims to question not only the direction American politics will take as it enters the twenty-first century but also the direction American politics has already taken.

The series addresses the role of interest groups and social and political movements; openness in American politics; important developments in institutions such as the executive, legislative, and judicial branches at all levels of government as well as the bureaucracies thus created; the changing behavior of politicians and political parties; the role of public opinion; and the functioning of mass media. Because problems drive politics, the series also examines important policy issues in both domestic and foreign affairs.

The series welcomes all theoretical perspectives, methodologies, and types of evidence that answer important questions about trends in American politics.

Rushed to Judgment

TALK RADIO, PERSUASION, AND AMERICAN POLITICAL BEHAVIOR

David C. Barker

COLUMBIA UNIVERSITY PRESS
NEW YORK

COLUMBIA UNIVERSITY PRESS

Publishers Since 1893

New York Chichester, West Sussex

Library of Congress Cataloging-in-Publication Data

Barker, David C. (David Christopher)

Rushed to judgment? : talk radio, persuasion, and American political behavior / David C. Barker

p. cm. — (Power, conflict, and democracy)

Includes bibliographical references and index.

ISBN 0–231–11806–6 (cloth : alk. paper) — ISBN 0–231–11807–4 (paper : alk. paper)

1. Talk shows—United States. 2. Radio in politics—United States. I. Title. II. Series.

PN1991.8.T35 B37 2002

791.44'6—dc21

2002019243

Columbia University Press books are printed on permanent and durable acid-free paper.

Printed in the United States of America

c 10 9 8 7 6 5 4 3 2 1

p 10 9 8 7 6 5 4 3 2 1

This book is dedicated to my mother, Dorothy, and to my daughter, Courtney

Contents

List of Figures

List of Tables

Acknowledgments

First and foremost, I would like to thank my mentor and friend, Kathleen Knight, for inspiring me and pushing me on this project from the beginning, and for providing invaluable training. As many others can attest, no one has ever benefited from more conscientious guidance than that given by Professor Knight to her students. Kathleen, I cannot thank you enough for questioning me, encouraging me, listening to me, and making me laugh.

Second, I would like to thank Christopher Carman for being a tough, honest, and tireless friend to me throughout the years that I have spent writing this book. Several of the ideas presented in this book originated and crystallized through daily conversations in the car to and from the office, or on the plane to and from professional conferences. Oh, and he makes a pretty good "best man" too.

Third, I would like to thank several people who have read different drafts or sections of this book over the years, and have made invaluable comments, without which this never would have seen the printed page: Robert Erikson, Mark Franklin, Christopher Wlezien, Carrie Funk, Lee Sigelman, John Zaller, Vanessa Baird, Brett Kleitz, Rick Matland, Doris Graber, Lynda Lee Kaid, Diana Owen, Vincent Price, Robert Lineberry and of course the anonymous reviewers for Columbia University Press.

Fourth, I would like to thank my colleagues at the University of Pittsburgh for supporting me both materially and emotionally while I have completed this project. In particular, Jon Hurwitz, Ray Owen, and Susan Hansen have provided invaluable advice and encouragement. I also thank Scott Beach and the staff at University Center for Social and Urban Research Survey Research Center for conducting an excellent survey of Allegheny County partisans that serves as the data for chapter five.

Fifth, I am indebted to Robert Shapiro as the editor of the Power, Conflict and Democracy series in American Politics for choosing to support this project, and to John Michel and the rest of the people at Columbia University Press for editing and publishing this book, returning my phone calls and emails, checking up on me, and providing a very professional and through review process.

Sixth, I would like to thank the National Science Foundation for supporting this research with a dissertation grant, which enabled me to conduct the experiments that serve as data in chapter 3, and for supporting the National Election Studies, which provide much of the data in other chapters.

Personally, I could not have finished this book without the inspiration and wisdom of my daughter, Courtney Barker. She reminds me every day that family, baseball, and pizza are more important than political science. Neither can I ever express enough gratitude to my mother, Dorothy Notgrass for her unwavering support, love, guidance, and example throughout my life. My stepfather, Troy Notgrass has also taught me a great deal, supported me, and provided a good sounding board for many of my ideas. Nor are there enough thanks to be given my wife, Susie, for making me feel loved every day and for great daily conversations (debates?) pertaining to media, politics, and religion, among other things. I love you all very much. Finally, I thank Bill Rushing, Gary Dickey, Jason Reichert, and Chuck Williams for their friendship, good humor, and patience with a guy who rarely returns phone calls on time when he is obsessing over a book deadline.

Last and least, I would like to thank Rush Limbaugh, whose impervious immunity to truth in broadcasting during the summer and fall of 1993 (when I spent many daytime hours in the car) provided the initial impetus for my scholarly interest in the political effects of talk radio.

Rushed
to Judgment

1 Introduction

We Americans are changing the way we gather political information. Perhaps due to our increased access to information and due to changes in lifestyle, we increasingly seek information that can be obtained conveniently, that doubles as entertainment, or that provides a perspective with which we sympathize. Thus while millions of Americans still peruse a daily newspaper and/or religiously view the evening network news, millions more bookmark their preferred political websites, watch political news magazines on cable television, or tune in to talk radio during their daily commutes. Such growth in usage of "new media" (Davis and Owen 1998) may have substantial implications for democratic discourse in the "marketplace of ideas." While the traditional media (e.g., newspapers, TV news, and major news magazines such as *Time* and *Newsweek*) attempt to uphold occupational norms of objectivity and equal time in their coverage of political events (Bennett 1988),[1] the new media are not regulated by such canons. Therefore as more Americans receive information from sources whose primary objectives are to entertain and persuade, democratic dialogue may become more misinformed, contentious, and polarized—resulting in legislative gridlock and/or restricted policy alternatives.

For political scientists, social psychologists, and communication scholars, the new media may provide fresh opportunities to find evidence of persuasive media influence over audience members' beliefs, opinions, and behavior. However, what analysis of the new media offers on one hand in terms of new opportunity, it takes away with the other, for the new media invite a notoriously self-selected audience. Cognitive dissonance theory posits that individuals may avoid messages that they find potentially distasteful, relying entirely on sources that appear kindred in spirit (Festinger 1957). Thus untangling the reciprocal causality between audience exposure to new media and political behavior poses a heavy analytical challenge.

Political talk radio typifies the new media. Convenient, entertaining, and provocative in its discourse, most political talk radio is unapologetically ideological in message. This book attempts to systematically and comprehensively examine the manner and extent to which listening to this popular medium may result in persuasion, broadly defined. Whereas persuasion may occur via a number of different processes, both within and beyond the context of talk radio, I apply a framework that conjoins framing and priming theory to explore how talk radio listening may make some considerations more accessible in memory, thus manipulating the relative salience of competing considerations as determinants of belief and choice (Zaller 1992; Nelson, Clawson, and Oxley 1997).

In examining the effects of persuasive efforts by talk radio hosts, much of my applied focus is on the medium's ringleader—Rush Limbaugh. With the most identifiable message content and by far the largest audience of any political talk host, Limbaugh provides a straightforward opportunity for assessing talk radio as a vehicle for political persuasion.

Substantively, this book first considers the manner in which persuasion via talk radio may occur, using carefully controlled experiments to assess causality in a way that broad quasi-experimental designs cannot. In so doing, I examine whether there is any basis for expecting real-world effects and provide one model of how those effects might happen. Operationally, I break down the Limbaugh message into two distinct yet broad propaganda techniques: *rhetoric*, or the attempt to persuade by offering new information (which may be either rationally or emotionally charged), and *value heresthetic*,[2] which prompts listeners to think in terms of higher-order values or principles, framing the issue in question around a particular value dimension, thus manipulating the salience of information already stored in memory.

After considering *how* opinion may be induced by talk radio hosts, this book goes beyond the lab to examine the *extent* to which talk radio listening is associated with opinion, activity, and belief. Chapters 4, 5, and the first part of chapter 6 focus on the relationship between what Rush Limbaugh says and the way listeners think or behave, measuring the persuasive utility of Limbaugh's best efforts. The second part of chapter 6 and all of chapter 7 shift gears, contemplating whether characteristics of the medium more generally have consequences in terms of the political realities that audience members construct.

Therefore this book attempts to shed light on a number of theoretical and applied puzzles. First and foremost, this book seeks to understand how political persuasion occurs. If politics is the authoritative allocation of value (Easton 1965), or "who gets what, when and how" (Lasswell 1958), or the pursuit, organization, and consequences of *power*, then democratic politics is fundamentally about persuasion. How is power achieved in a democracy if not by persuading others to buy your "product" within the "marketplace of ideas?"

Second, this book seeks to understand, in as thorough a way as possible, what the possible effects are of one of the most conspicuous forms of new media: call-in talk radio. Some have asked whether this new medium can serve as an agent of deliberative democracy, spurring Americans to form pseudocommunities, where policy choices are debated in an open forum, thus bringing American politics closer to a democratic ideal (Page 1996). Others wonder whether talk radio has rekindled the partisan press of times past, when objective journalism was jettisoned for polemic. Now is a good time to evaluate the effects of political talk radio because the medium is no longer a fad and shows no signs of fading into oblivion. Furthermore, talk radio serves as a poster child for the new media—unabashedly subjective, entertaining, ubiquitous, and convenient. Given the prominent place of talk radio stations in most markets' AM dials, we have now had the time to critically evaluate the effects of a medium that is not disappearing. Indeed, many young conservatives who do not remember Ronald Reagan have "grown up" with Rush Limbaugh. Does listening to talk radio change the way people think about politics, or are attitudes on the part of listeners purely a function of the nature of the audience? Does it inspire people to be more-active and more-committed democrats, or does it lead to cynicism and distrust? Does it enhance public understanding of public issues, or serve as a breeding ground for greater misunderstanding? How does this affect the way we talk and relate to each other—the quality and civility of discourse? Is there any turning back?

The remainder of this chapter introduces and summarizes the extant political communication literature as it pertains to political persuasion, propaganda, and media effects, laying a theoretical foundation for the substantive chapters that follow. Chapter 2 continues this effort, focusing specifically on talk radio—its scope, audience profile, and main message characteristics.

Political Persuasion, Propaganda, and Media Effects

At its root, politics is about how you "get people who start off on one side of the room to move to another" (Sniderman 1993). Whether in the form of the president "going public" (Kernell 1986), candidates standing before the camera in a thirty-second advertisement, or one voter trying to induce another to vote for the Democrats this time, persuasion lubricates democratic process (Mutz et al. 1996; Barber 1984; Dahl 1989; Fishkin 1995).

For the purposes of this book, I interpret persuasion broadly, to include *any inducement of the beliefs, attitudes, or choices of an individual or collective body by another. Beliefs* are what an individual considers to represent objective information, or "truth." *Attitudes* are somewhat stable orientations of affect toward some object, person, or idea. *Choices* may include momentary opinions, policy preferences, candidate appraisals, vote selections, and participation decisions, among others.

Although the relative stability of attitudes may make them more difficult to manipulate than momentary choices, attitudes are no more relevant to democracy than the perhaps more-ephemeral choices that are reflected in public-opinion surveys, given the centrality of survey opinion to modern campaigning and governing (Zaller and Feldman 1992; Morris 1998). As such, this book concerns itself with persuasion, whether fleeting or persistent.

Persuasion has been of interest to scholars from a variety of disciplines since ancient Greece. In *Rhetoric* and *Topics*, Aristotle lectured on rhetorical devices—how to employ them for greater persuasive effect and how to guard against being manipulated by them. But it was Walter Lippmann's seminal *Public Opinion* (1922) and *The Phantom Public* (1925) that kicked off modern efforts to understand the interplay between mass communications and individual choices. Lippmann wrote that individual opinions are "pieced together out of what others have reported and what we can imagine" (Lippmann 1922:53), and went on to argue that *what others report* determines, to some extent, what we can imagine.

Two world wars and a perceived communist threat prompted concerns about clandestine attempts to manipulate the ideas and opinions of the public through words. As a consequence, early empirical work focused on content-analyzing political messages to chronicle the devices of propaganda (e.g., Lasswell, Casey, and Smith 1935). In *The Fine Art of Propaganda* (1939), the Institute for Propaganda Analysis (IPA) described several notorious propaganda devices, such as name calling, testimonials, bandwagon appeals, and "card stacking"—the dispositional arranging of evidence in a particular way to advance an argument. Over the years, others refined and expanded the IPA's efforts (e.g., Chase 1956).

But if communication research examines "who says what, to whom, with what effect" (Lasswell 1958), chronicling the incidence of propaganda in political messages did little to advance understanding of "what effect" messages have. Empirical exploration of persuasion *effects* exploded after World War II when Carl Hovland and his colleagues at Yale began systematically analyzing how persuasion occurs. The research produced by the Yale group remains definitive for many topics (e.g., Hovland, Janis, and Kelley 1953; Hovland and Rosenberg 1960; Sherif and Hovland 1961) and spawned volumes of work on the variables that influence persuasion and the processes through which those variables operate (e.g., Chaiken 1986; Petty and Cacioppo 1981).

The collective literature has concluded that there are two primary routes to persuasion. The first path, alternately labeled the "central" or "systematic" route, requires audience members to expend considerable amounts of cognitive energy. Audience members carefully and systematically consider the arguments placed to them and go through a process of self-suasion before making a choice that is perhaps in line with that of the message source. By contrast, the "peripheral" or "heuristic" route to persuasion requires relatively little mental effort on the part of the audience member. Audience members shirk cognitive responsibilities, relying instead on cognitive shortcuts (a.k.a. "heuristics") to make up their minds. Some of the heuristics upon which people most often rely include emotions, party identification, social desirability, or core values. Peripheral persuasion processes encourage people to rely on heuristics and take advantage of our natural propensity toward being "cognitive misers" (Popkin 1991; Sniderman et al. 1991).

Neither route operates for all people all the time. But all things being equal, people tend to "satisfice" when confronted with political decisions. That is, we stop working when we reach an acceptable, but not necessarily optimal, level of understanding (Kinder 1998). And why not? As Lippmann pointed out, to expect ordinary people to become enthralled with public affairs would be to demand of them an almost pathological affinity for politics. No, Americans are "much more concerned with the business of buying and selling, earning and disposing of things, than they are with the 'idle' talk of politics" (Lane 1962:25). In the great circus of life, politics is but a "sideshow" (Dahl 1961:305). Therefore, controlling for contextual and audience characteristics, the messages of propagandists who attempt to persuade via the central or systematic route may fall on deaf ears more often than not. As will be explained in the pages that follow, this book posits that successful persuasion on the part of talk radio hosts depends to some extent on their traversing heuristic, rather than systematic, routes to persuasion.

Persuasion Variables

Four main categories of variables influence whether and how persuasion will occur: source, recipient, context, and message characteristics. *Source* variables refer to individual aspects of the message sender(s). Some of the source variables that have been shown to have a significant impact on persuasion are credibility (Hovland and Weiss 1951), including expertise (Petty and Cacioppo 1981) and trustworthiness (Hass 1981); physical attractiveness (Snyder and Rothbart 1971); likableness (Chaiken 1986); perceptions of source power or position (McGuire 1969); speed of speech (Miller et al. 1976); gender (Goldberg 1968); majority/minority status (Asch 1956); and similarity to the receiver, either real (Brock 1965) or perceived (Lupia and McCubbins 1998).

Recipient variables refer to specific characteristics of the message receiver(s). Some of the recipient variables that have been shown to mediate persuasion include attitudinal variables—such as whether attitudes are strong (Petty and Krosnick 1995), how accessible the attitude is in memory (Jamieson and Zanna 1989), and issue-relevant knowledge (Wood, Rhodes, and Biek 1995); demographic variables, such as gender (Cooper 1979) and age (Alwin, Cohen, and Newcomb 1991); intelligence (Rhodes and Wood 1992); self-esteem (McGuire 1968); sensitivity to social cues (DeBono 1987); need for cognition (Cacioppo, Petty, and Morris 1983); and mood (Petty et al. 1993).

Context refers to factors that involve the setting in which the communication takes place. Context variables that have been shown to affect the persuasion process include distractions (Festinger and Macoby 1964), audience reactions to the source (Petty and Brock 1976), forewarning of the source's position (Freedman and Sears 1965), forewarning of persuasive intent (Hass and Grady 1975), anticipated discussion or interaction (Cialdini et al. 1976), and message modality (Chaiken and Eagly 1976).

Message characteristics refer to aspects of the communication itself. They are perhaps the most theoretically interesting variables, because these are the most readily controllable by a message sender. Among the message-content variables that have been studied extensively are the quantity of the information presented (Petty and Cacioppo 1984), the presence of a causal explanation within an argument (Slusher and Anderson 1996), the degree to which the consequences presented within an argument are likely and desirable (Areni and Lutz 1988), the positivity/negativity of an argument (Meyerowitz and Chaiken 1987; Cobb and Kuklinski 1997), the degree of emotion versus reason in an argument (Olson and Zanna 1993), whether strongest arguments are placed at the beginning or the end of a message (Haugvelt and Wegener 1994), whether arguments are simple or complex (Cobb and Kuklinksi 1977, whether arguments are one sided or two sided (Allen 1991), and how consequences of a proposed policy are interpreted (Lau, Smith, and Fiske 1991).

However, although argument quality is one of the most manipulated variables in the contemporary social psychological literature on persuasion, relatively little is known about what makes a message persuasive (Petty and Wegener 1998). Perhaps this is related to the psychologists' preference for considering arguments strictly as messages that try to change a recipient's mind by presenting information that, it is argued, makes some consequence likely to occur (Petty and Wegener 1998). But such a focus primarily explores how the central or systematic route to persuasion is achieved and perhaps fails to consider the ways that message variables influence peripheral or heuristic routes. As already noted, heuristic processes rely on cognitive shortcuts for decision making. This means applying inferential reasoning to draw conclusions about what is unknown from what is known. Hence persuasion that occurs via the heuristic route does not succeed by making audience members believe something new; it merely tries to make some already-held beliefs or prejudices more salient than others.

Media Effects

"Media effects" research within political science explicitly considers several heuristic routes to persuasion. Until recently, this literature was dominated by controversy over whether mass media could make a significant impact on the public at all. As already noted, propaganda analysis dominated early studies in political communication, and reflected the *hypodermic needle* model of media effects. This perspective assumes that media are able to inject or otherwise infect audiences with ideas. It considers the public to be a more or less undifferentiated mass, vulnerable to powerful media that intend to indoctrinate (e.g., Charters 1933). Indeed, folklore is replete with widely believed stories of media influence, from William Randolph Hearst's alleged hand in the Spanish-American War to Richard Nixon's sweating debate performance and his narrow defeat in the presidential race of 1960.

However, researchers have been hard pressed to find more than anecdotal evidence of straightforward media domination. In light of a series of disappointing efforts to demonstrate the media's power, hypodermic theories gave way to the *minimal-effects* verdict (Berelson, Lazarsfeld, and McPhee 1954; Klapper 1960). The minimal-effects perspective contends that a number of mediating conditions prevent the media from having a significant impact on audiences. The most salient of these conditions involves *selection bias*.

Selection bias contains three elements: (1) selective exposure—the inclination of people to expose themselves mainly to the media content that they expect will be compatible with their views; (2) selective perception—the biased processing that people employ when they encounter a message, perceiving it in accordance with their preexisting beliefs and attitudes; and (3) selective retention, when people disproportionately remember information that is consistent with what they already believe or prefer, even if they have to distort that information somewhat.

Selection bias arguments have their roots in cognitive-dissonance theory (Festinger 1957). Two elements in a cognitive system (e.g., two attitudes, a belief and an attitude, an attitude and a behavior) are said to be dissonant if they imply the opposite of each other. For example, if an individual strongly believes in civil liberties and the right to privacy, yet for other reasons maintains a pro-life position on abortion, the resulting ambivalence might cause that individual to experience cognitive dissonance—confusion, self-doubt, and perhaps guilt. Such dissonance, it is claimed, leads individuals to avoid exposing themselves to or retaining information that might engender such cognitive discomfort. As Joseph Goebbels, the notorious minister of propaganda during the reign of the Third Reich, once noted, "There is nothing that the masses hate more than two sidedness, to be called upon to consider this as well as that. They want to generalize complicated situations and draw uncompromising solutions" (Lochner 1970).

But there are a number of reasons to be skeptical of the minimal-effects conclusion. First, with regard to selection bias, dissonance research has shown that many people do not have a need for cognitive consistency (Cialdini, Trost, and Newsom 1995). Moreover, cognitive inconsistency per se may not lead to discomfort; discomfort may depend upon whether people believe that the dissonance threatens their moral integrity (Cooper and Fazio 1984; Steele 1988).

Furthermore, social judgment theory predicts that persuasion is an increasing function of an audience member's *latitude of acceptance, rejection, or noncommitment* (e.g., Sherif and Hovland 1961). In other words, in order for persuasion to occur, messages must be somewhat discrepant to those already held by recipients. The theory posits that everyone has a range of attitudes that they might take in a given situation, not just a precise, pinpointed opinion. As long as a message source is not expected to espouse ideas that are known to be squarely outside an audience member's latitude of acceptance or noncommitment, then selection bias should not contaminate the makeup of the audience. This theory has been supported by a variety of studies (e.g., Hovland, Harvey, and Sherif 1954; Aranson and Carlsmith 1963). As Nelson and his colleagues describe it, "our experiments reveal that even relatively knowledgeable people do not necessarily have fixed opinions on matters of public debate. For most people, attitudes on most political issues are not like files in a drawer, waiting to be pulled out and consulted whenever the need arises. Rather, 'attitude' should properly refer to a range of potential evaluative expressions" (Nelson, Clawson, and Oxley 1997:237).

Perhaps even more important, Zaller's (1992) RAS (reception acceptance sampling) model of public opinion convincingly demonstrates that individuals selectively avoid certain messages only to the extent that they are sophisticated enough (in terms of political knowledge, ideological development, and so on) to recognize that those messages are discrepant with their beliefs. But a great deal of evidence has now been collected to conclude that most individuals know very little about politics, care even less, and are for the most part "innocent of ideology" (Delli Carpini and Keeter 1996; Converse 1964; Kinder 1983). Simply stated, "Most Americans glance at the political arena bewildered by ideological concepts, lacking a consistent outlook on public policy, and in possession of genuine opinions on only a few issues" (Kinder 1998:794).

Another reason to be wary of the minimal-effects conclusion is that, as the *uses and gratifications* approach (e.g., Lull 1990) and the *media systems dependency* approach (De Fleur and Ball-Rokeach 1988) dictate, people select media content for a variety of reasons, including (perhaps primarily) its entertainment value. The ideological tenor of program content, while perhaps an important selection criterion for some people, constitutes but one reason, among many, for choosing among media alternatives.

Heresthetic

One explanation for early failures at finding meaningful media influence may relate to a preoccupation with finding persuasion via the central route—persuasion that occurs as a result of learning new information. Indeed, "political elites may fail to influence public opinion among the most knowledgeable through direct propaganda campaigns, but they may succeed in directing public opinion in their favor through clever frames" (Nelson, Clawson, and Oxley 1997:239). In other words, beyond trying to persuade by directly spreading information or misinformation, and thus perhaps prompting message receivers to counterargue, a message source may present an issue in such a way as to take advantage of the message receiver's ambivalence toward the issue (Hochschild 1981), making some considerations seem more important than others and thus promoting a particular problem definition, causal interpretation, moral evaluation, and/or treatment recommendation (Entman 1993).

This *framing* concept can be traced to Schattschneider, who argued that "political conflict is not like an intercollegiate debate in which the opponents agree in advance on a definition of the issues. As a matter of fact, *the definition of the alternatives is the supreme instrument of power*; the antagonists can rarely agree on what the issues are because power is involved in the definition. He who determines what politics is about runs the country" (1960:68; italics in original). Schattschneider's bold contention has now been supported empirically across many contexts, demonstrating how the media and others may persuade via the peripheral route by framing (Dorman and Livingston 1994; Gamson and Modigliani 1989; Gitlin 1980; Iyengar 1991; Nelson and Kinder 1996; Pan and Kosicki 1993; Patterson 1993; Nelson, Clawson, and Oxley 1997).

So frames, without necessarily providing any new information about an issue, tell people how to weigh the often conflicting considerations that enter into political deliberations. But by what means can a message source manipulate the relative salience of competing considerations? One way may be to simply make some considerations more accessible in memory. *Priming theory* (e.g., Iyengar and Kinder 1987; Krosnick and Kinder 1990; Tourangeau and Rasinski 1988; Zaller and Feldman 1992) suggests that when people are faced with making a political decision, they are not cognitively capable of bringing to bear everything they believe about that issue. As such, their decisions will be a product of whatever ideas happen to be most accessible in memory, or at the "top of their head" (Zaller 1992). By framing issues in one way or another, media and other sources may be able to prime some considerations to be more accessible. The considerations that are most accessible are also the ones most likely to be thought of as most important by the message recipient. Therefore while *framing*, or the manipulation of consideration importance, may occur via other more-deliberate processes as well, *priming*, or the manipulation of consideration *accessibility*, provides one well-traversed path. As such, while framing and priming enjoy distinct identities, they can be compared to Siamese twins, in that they are difficult to separate. For this reason, as mentioned earlier, when describing this joint process of priming and framing, I will borrow Riker's term, *heresthetic*, defined here as *the strategic redefinition of an issue by manipulation of the salience (accessibility and importance) of considerations through framing and priming*.

Heresthetic theories also differ from traditional rhetorical theories of persuasion in the way these processes interact with audience sophistication. Traditional persuasion models have shown that, assuming a message is received and understood equally well by both sophisticated and unsophisticated audiences, the more sophisticated are less likely to be persuaded by or "accept" that message, because they are more likely to already hold an opinion on that issue, and are more likely to have sufficient understanding of the issue. Armed with such intellectual capital, sophisticated audience members are more likely to reject the persuasive intent (Eagly and Chaiken 1993; McGuire 1968, 1985; Zaller 1992).

However, differences in sophistication should not depress heresthetic effects. Because such effects do not depend upon a recipient's acceptance of a message's particular claims—instead operating by calling to mind considerations already stockpiled in memory—sophisticated respondents should, if anything, be *more* susceptible to framing/priming effects. Research by Nelson and colleagues (1997) provides compelling evidence that this is indeed the case.

So this book considers how political persuasion occurs, through central and peripheral routes, via heresthetic or rhetoric, within the context of one of the most salient examples of the "new media"—call-in political talk radio. I consider such persuasion processes theoretically and empirically and examine persuasion in many forms, including opinion inducement, belief change, value priming, and mobilization. I examine various forms of political behavior including knowledge, opinion, turnout, partisanship, proselytizing, and vote choice—across presidential, congressional, and primary elections. I seek to understand both whether persuasion occurs and how—both whether talk radio matters and why.

Specifically, chapter 3 describes the execution and results from an experiment designed to assess how, when pitted head to head, value heresthetic stacks up to rhetoric as a persuasion determinant, using talk radio host Rush Limbaugh as the message source. Chapter 4 moves beyond the lab to a straightforward examination of the relationship between Limbaugh listening and political preferences, applying several techniques to combat the threat of selection bias. Chapter 5 extends this analysis, applying it to the 2000 Republican nomination struggle between John McCain and George W. Bush.

Chapter 6 considers persuasion in terms of political mobilization. Building upon the well-grounded relationship between internal political efficacy and political participation (Abramson and Aldrich 1982), the chapter asks whether individual levels of political efficacy are affected by the media messages to which one is exposed, thus increasing one's likelihood of participating in politics. The first part of the chapter expands the previous heresthetic framework, extending the experimental design of chapter 3 to evaluate the viability of Limbaugh messages that are designed to make listeners feel more efficacious—whether that increased efficacy is manifested in greater participation in simulated legislative committee deliberations.

The Construction of Political Meaning

The second part of chapter 6 considers the talk radio audience as a pseudocommunity, applying theories from social context studies (e.g., Huckfeldt and Sprague 1987) to see if conservative listeners become more efficacious and participatory after listening to conservative call-in programming, having constructed a political reality from the programming that perceives conservatism to be the dominant belief structure among "the people." Conversely, I test to see if liberal listeners become less efficacious in response to this constructed message, thus falling into a "spiral of silence" (Noelle-Neumann 1984).

This second portion of chapter 6 moves away from the heresthetic model of political persuasion on which the earlier chapters are based (which considers the influence of specific political messages) to a theoretical foundation in constructionist theory. Constructionism contemplates how individuals actively construct (a perhaps tinted) political reality from the limited range of messages to which they attend (Crigler 1996). In the context of conservative talk radio, based on the barrage of conservative callers on the Limbaugh show, constructionism may mean drawing the inference that there exists a greater amount of political conservatism among the electorate than empirics would reveal is warranted. Depending upon a listener's ideological bent, such constructed reality may make a listener either more or less efficacious—and thus more or less likely to participate in politics.

Chapter 7 continues to explore talk radio priming from a constructionist perspective, examining how talk radio may make some beliefs (perceived knowledge as opposed to opinion) more accessible by its tone and content. Listeners may apply inferential reasoning while listening, coming to confidently believe political falsities even though the talk radio messages may not have overtly spread such mistruth.

So in sum, this book seeks to understand how audience members react to media messages whose primary aim is not objective journalism. Combining framing and priming theories under the rubric of heresthetic, I first consider how the substance of such media messages may induce attitudinal and behavioral changes. Second, relying on constructionist theory, I explore the democratic consequences, in terms of participation and information, of widespread exposure to a medium grounded in the principle of expanded democratic dialogue. Therefore this book examines democratic politics in terms of political discourse. The particular lens through which I view the ramifications of such discourse involves the applied question of talk radio influence. But before I address this question, I must describe the medium, its audience, and the characteristics of its message. It is to this task that I turn in chapter 2.

2 Political Talk Radio and Its Most Prominent Practitioner

Chapter 1 reviewed the exhaustive literature in political communication that deals with persuasion, emphasizing media effects. A large body of research now points to the conclusion that media effects are more "fugitive" than minimal—meaning they are out there, just hard to find (e.g., Bartels 1993; Page, Shapiro, and Dempsey 1987; Dalton, Beck, and Huckfeldt 1998). The search has often been confounded by reliability and validity challenges. Not only should we raise a suspicious eyebrow toward self-reports of media exposure, because social desirability encourages survey respondents to inflate the attention they pay to political news (Weisberg, Bowen, and Krosnick 1989), but most studies have neglected (or have been unable) to examine the specific media content to which research subjects have been exposed, relying instead on measures of how often survey respondents watch television news, for example. Perhaps even more important than the striking loss of efficiency associated with the use of such error-laden measures, which may have accounted for many minimal-effects conclusions (Bartels 1993), is the strong possibility that media effects are often not observed in the aggregate, because partisan or ideological messages often counterbalance each other in the traditional mass-media universe. For instance, busi-

---

Parts of this chapter appear in "Talk Radio Turns the Tide? Political Talk Radio and Public Opinion" (Barker and Knight 2000). Used by permission.

ness enthusiasts may read the *Wall Street Journal*, exposing themselves to political news with a pro-business or conservative slant. On the other hand, those with a particular interest in international affairs might choose the *New York Times* for its extensive international coverage, all the while taking in an editorial page that some would say is sympathetic to Democrats. Furthermore, ideological points of view might even balance out within a single publication—Maureen Dowd and William Safire might cancel each other out in the editorial pages of many American newspapers, for example. Moreover, local media consumption may pose even bigger challenges to determining exactly what messages the consumers are getting.

Large media effects are more likely to be observed when news coverage is particularly one-sided, such as the coverage of the Gulf war (Price and Zaller 1993), and the state of the economy during the 1992 electoral campaign (Hetherington 1996). As such, political talk radio provides an ideal medium through which to assess media influence. This chapter begins with a general discussion of political talk radio: its audience, content, media characterization, public evaluation, and scholarly treatment. From there, I move to a strict emphasis on the message of Rush Limbaugh—the issues he highlights, the positions he takes, and the persuasion techniques he employs.

Background

Political talk radio may be defined as *radio programs (usually sporting a call-in format) that emphasize the discussion of elections, policy issues, and other public affairs*. Originating in the 1930s, talk radio was a popular outlet for politicians. Of course, Franklin Roosevelt's fireside chats during the Great Depression are legendary. Roosevelt's radio speeches during the 1940 campaign were heard by as much as 39 percent of households owning radio sets (Chase 1942). But political talk radio has never been the exclusive domain of those holding or running for office. Perhaps as a counterweight to Roosevelt, Father Charles Coughlin held an audience of approximately ten million for his weekly broadcasts attacking the New Deal (Tull 1965; Brinkley 1982).

Nationally syndicated radio call-in programs had their genesis in the 1970s. The Larry King Show was the most prominent of these early shows, with more than three hundred affiliates. But political discussion often took a back seat on these shows to entertainment personalities, popular psychology, and the like. Political talk radio as we know it today began in the 1980s and has flourished in the 1990s. (Capella, Turow, and Jamieson 1996; Davis and Owen 1998).

But despite its popularity, talk radio is very controversial. While some hail it as America's new "back fence," fostering pseudocommunities and providing the ultimate arena for free, democratic discourse (Ratner 1995; Levin 1987), others worry that talk radio may foster listeners' basest instincts. Critics complain that we have reached a point where ignorant blather is considered on par with informed commentary (e.g., Dreier and Middleton, 1994), where the views of Henry Kissinger carry only a slightly higher price tag in the political marketplace than those of "Joe, from Round Rock, Texas."

One of the events that paved the way for the success of political talk radio was the Federal Communication Commission's decision in 1985 that the Fairness Doctrine was no longer needed, a decision that was unsuccessfully challenged by Congress and subsequently upheld by a federal Appeals Court in 1989. Adopted in 1949, the Fairness Doctrine had stipulated that broadcasters must provide reasonable balance when airing controversial opinions. With the end of the Fairness Doctrine, broadcasters were free to air ideologically biased programming.

Indeed, as opposed to traditional media, political talk shows are unabashedly biased. While some networks, such as ABC, Major Radio Network, and Westwood One carry both liberal and conservative hosts, and the small Pacifica network is an example of a leftist network, the majority of political talk programs feature conservative, libertarian, or populist hosts (Davis and Owen 1998). In response to challenges of unfairness, some hosts contend that rather than stifling opposing viewpoints, their existence provides needed balance to the mainstream media. Host Blanquitta Cullum argues that "by using the evening news as a left-wing doormat, they [liberals] have created demand for a right-wing product" (Cullum 1994). Similarly, Rush Limbaugh says ". . . my views and commentary don't need to be balanced by equal time. I am equal time. And the free market has proven my contentions" (Limbaugh 1994).

Hosts also claim that callers offer a different range of perspectives. Limbaugh often brags, "liberals are pushed to the front of the line." But balance is not the goal of such devices. Political talk radio must appeal to the marketplace; hence the primary goal of most shows is entertainment. An open-minded consideration of the various sides of an issue is not entertainment to most listeners. By contrast, verbal conflict that culminates with a clear "winner" spurs interest. As Limbaugh admits, "the primary purpose of a call is to make me look good, not to provide a forum for the public to make speeches" (Limbaugh 1992).

Format

Talk radio programs typically conform to the following script: an opening monologue by the host, sometimes followed by the introduction of a guest or guests, accompanied by interaction between the guest or host and callers. With or without guests, the host is the headliner of the program. Davis and Owen (1998) remark that the host is more like Geraldo Rivera or Oprah Winfrey than Dan Rather or Judy Woodruff. The host's opening monologue carries tremendous import. It sets the tone and agenda for the remainder of the show, establishing the topics that will be up for discussion and the host's (never retreating) position on those issues. Opening monologues can last anywhere from just a few minutes up to half an hour, depending on the host's interests, the presence of guests, and the news cycle. Some hosts supplement opening monologues with shorter monologues at the beginning of each hour. With ten to twenty hours of airtime to fill each week, hosts have time to discuss issues at length. However, just because talk radio hosts have more time to delve into issues does not necessarily translate into a more substantial treatment of those issues than would be found in the mainstream press. Programs often cover a wide range of issues—moving quickly between callers, host pontification, and advertising—which often translates into superficial discussion (Davis and Owen 1998).

Following the opening monologue, the host usually begins taking calls. A call screener answers the phone and conveys information about the caller to the host by typing onto a terminal that the host can read through a computer screen. That information includes the caller's name, approximate age, gender, location, and a brief summary of the point that the caller wants to make. Thus the screener is a filter whose job is to enhance the appeal of the program. That usually means limiting calls from those over fifty years of age, those who cannot effectively articulate their point, or those who are likely to make the host look bad (Davis and Owen 1998). Studies have also found that male callers are more than twice as likely as female callers to achieve airtime.

The majority of callers usually agree with the host, at least when the host is conservative (Davis and Owen 1998). This may be because more people who agree are likely to call, or it may be attributable to the screening process. Conflict spurs interest, so it stands to reason that screeners would want to "put through" callers who disagree with the host. However, as Limbaugh has noted, the primary purpose of a call is to glorify the host. As a consequence, *skilled* callers who disagree may be screened out. Once on the air, callers have a limited amount of time to make their point. Rarely does a host interact with a caller for more than two or three minutes. Women and the elderly, if they get through, are often allotted even less time (Davis and Owen 1998).

Media Portrayal

What is the nature of mainstream media attention to political talk radio? A large-scale content analysis by scholars at the Annenberg School of Communication (Capella, Turow, and Jamieson 1996) found that, in general, the print media's attention to talk radio is narrow and unfavorable. First, the Annenberg scholars' study indicates that a reader of the mainstream print media would find little in-depth investigation of talk radio hosts or their programs. In fact, "the mainstream print media pay little attention to issues discussed on radio's political talk programs" (38). Furthermore, the analysis found that the press tend to describe talk radio as a pernicious force. Fewer than 5 percent of the articles included in the content analysis reflected any degree of positivity toward the mentioned talk radio host or show. Typical examples of print media characterization from the Annenberg analysis include the following:

> Listening in for a day is to be pelted with tales and travails, vehemence and vitriol, paranoia and pettiness, stupidity masquerading as wisdom and, occasionally even vice versa. (Weber 1992)
>
> There is a meanness in the land. We can hear it in the angry howls on talk radio. As Limbaugh or some other imitator goads his listeners on, the basic message is: we are entitled to our meanness. (Gabler 1995)
>
> What passes for political debate on many talk shows is often a cacophony of inflammatory rhetoric and half-truths. (Dreier and Middleton 1994)

Moreover, the press tend to portray talk radio as being homogeneous. Most articles that discuss talk radio content provide outrageous and disturbing quotations from Bob Grant or G. Gordon Liddy and then proceed to generalize such comments to talk radio more broadly. They rarely acknowledge that other, more-moderate forms not only exist but dominate the medium. This negative media portrayal may explain why nonlisteners tend to view political talk radio quite negatively (Capella, Turow, and Jamieson 1996).

Finally, the Annenberg scholars found that the press portray political talk radio not only as nefarious but also as a powerful force in American politics. They conclude that those reading story-length accounts of talk radio in the mainstream media would infer that talk radio has been extraordinarily effective in blocking or overturning legislative action, advocating legislation, influencing political behavior, and mobilizing political support (Capella, Turow, and Jamieson 1996).

Audience

The talk radio audience is considerable. The Annenberg survey found that 36 percent of the public listened to political talk radio at least occasionally in 1996, with 24 percent listening at least once a week and 18 percent listening at least twice a week.[1] A number of scholars and journalists have sought to paint the profile of this quarter of the eligible voting population that listens regularly to political talk radio. Early studies concluded that talk show listeners were older, less affluent, and less educated than nonlisteners (Crittenden 1971; Surlin 1986) and more socially isolated (Avery and Ellis 1979; Bierig and Dimmick 1979). Contemporary research, however, has found that the talk audience is *more* affluent and issue oriented than its nonlistening counterparts. In a sample of San Diego listeners, Gianos and Hofstetter (1995) concluded that *nonlisteners* tended to be less well educated, lower in income, and shorter-term residents than listeners. Listeners also tend to have more knowledge about civics and current political events than do nonlisteners (Davis and Owen 1998; Capella, Turow, and Jamieson 1996). In terms of demographics, listeners do still tend to be older and are more likely than nonlisteners to be white males who call themselves "born-again" Christians (Davis and Owen 1998; Capella, Turow, and Jamieson 1996).

Why do people tune in? Early studies found that motives associated with listening to talk radio include use of the medium as a surrogate companion (Avery, Ellis, and Glover, 1978), desire for entertainment, escapism, convenience, relaxation, and passing time (Armstrong and

Rubin 1989). However, contemporary research reports that the primary motivation for listening to talk radio today is information seeking. "Listeners want to keep abreast of issues, learn what others think, find out more about things they have heard about elsewhere, and provide reinforcement of their own political views. Entertainment, personal interest, and passing the time are also motivating factors for talk radio listeners" (Davis and Owen 1998:159).

Many have argued that talk radio is an expression of widespread alienation and discontent, a way of externalizing frustrations with politics and politicians (Bick 1988; Levin 1987). However, applying methodological rigor to this question, Hofstetter and colleagues (1994) found that talk radio listeners are not politically cynical and socially alienated. On the contrary, Hofstetter and colleagues found that political talk radio was associated with political involvement and activity. In the San Diego sample, frequent listeners to political talk radio were more interested in politics, paid more attention to politics in mass media, voted more, and participated more than others in a variety of political activities. Listeners were also more efficacious and less alienated than nonlisteners. These findings have been replicated in national surveys conducted by the Pew Research Center for the People and the Press (Davis and Owen 1998), the *American National Election Studies, 1995–97* (Barker 1998a), and the Annenberg School of Communication at the University of Pennsylvania (Capella, Turow, and Jamieson 1996).

Talk radio listeners are generally thought to be disproportionately conservative in their ideological identification. Surprisingly, however, Owen (1995) found that although "those who tune into talk radio tend to be slightly more Republican and Independent than their nonlistening counterparts, these differences are not statistically significant" (62). Gianos and Hofstetter (1995), in a sample of San Diego residents, affirmed that listeners are not simple sycophants suborned to the will of flamboyant hosts. The authors concluded that listeners expressed considerable disagreement with hosts, with only 4.9 percent of respondents reporting that they agreed with the talk show host nearly "all the time."

As table 2.1 displays, talk radio listeners are more likely to identify themselves as conservative than nonlisteners (Davis and Owen 1998). But ideological differences are not as dramatic as one might expect. As the 1996 American National Election Study (ANES) reports, although talk radio listeners are significantly more likely to call themselves conservative (57% to 42%), they are not significantly more inclined to agree that it is "not a problem if people don't have equal rights" (41%

Table 2.1 Opinions on Selected Issues and Attitudes Toward Government for Talk Radio, Television News Magazine, and On-line Media Audiences

	Audience (%)			
	General Public	Talk Radio	News Magazine	Online
Issue	**Not a Problem People Don't Have Equal Rights**			
Agree	37	41	43	30
Neutral	19	43	20	20
Disagree	44	39	37	50
Total	100	100	100	100
	Gone Too Far Pushing Equal Rights			
Agree	54	57	56	58
Neutral	14	13	15	16
Disagree	32	30	29	26
Total	100	100	100	100
	Best Not to Be Involved in Helping Others			
Agree	35	35	39	26
Neutral	17	17	20	18
Disagree	48	48	41	56
Total	100	100	100	100
	Fewer Problems More Traditional Families			
Agree	85	87	89	77
Neutral	08	06	07	11
Disagree	07	07	04	12
Total	100	100	100	100
	Newer Lifestyles Bad for Society			
Agree	70	73	76	61
Neutral	14	12	11	16
Disagree	16	15	13	23
Total	100	100	100	100

Table 2.1 Opinions on Selected Issues and Attitudes Toward Government for Talk Radio, Television News Magazine, and On-line Media Audiences *(continued)*

	Audience (%)			
	General Public	Talk Radio	News Magazine	Online
			Most Important Problem	
Good Job	07	07	10	06
Fair Job	44	39	52	37
Poor Job	49	54	39	57
Total	100	100	100	100
Less govt. better	45	57	46	53
Govt. should do more	55	43	54	47
Total	100	100	100	100
Need strong govt. to handle problems	62	49	63	58
Free market can handle problems without govt.	38	51	37	42
Total	100	100	100	100
Govt. bigger too involved	50	61	50	54
Govt. bigger	50	39	50	46
Total	100	100	100	100
Govt. wastes taxes				
A lot	60	63	65	58
Some	38	35	34	41
Not much	02	02	01	01
Total	100	100	100	100
Govt. run by few big interest	72	73	69	67
Govt. run for good of all	28	27	31	33
Total	100	100	100	100

Source: 1996 American National Election Study.
Previously published in Davis and Owen (1998:169, 176).

to 37%), that we "have gone too far pushing equal rights" (57% to 54%), that it is "best not to be involved in helping others" (35% to 35%), that there "would be fewer problems if we had more traditional families" (87% to 85%), or that "newer lifestyles are bad for society (73% to 70%).

On the other hand, with regard to political economics, talk radio listeners do appear to be substantially more conservative, as evidenced by their being more likely to agree that "the less government the better" (57% to 45%), "the free market can handle problems without government" (51% to 38%), and that "the government is bigger because it is too involved in the economy" (61% to 50%). However, listeners are no more likely to believe that "government wastes taxes a lot" (63% to 60%), or that "government is run by a few big interests" (73% to 72%).

Content

What do people hear when they tune in to political talk radio? Across the United States, political talk radio programs number in the hundreds, and they are anything but monolithic in format, style, or common subject matter. To understand the diversity of the content, Capella, Turow, and Jamieson (1996) conducted a careful content analysis of fifty of the most popular programs (twenty-four conservative hosts, seventeen moderate hosts, and twelve liberal hosts) for two weeks during the 1996 presidential primaries (4–15 March). These researchers found that moderate and conservative shows tend to cover foreign affairs and military matters at a higher rate than liberal shows, and that liberal shows tend to give more attention to issues involving education, children, prayer, gender roles, and ethics. Crime, courts, and justice appear to receive the most attention from moderate hosts. But the differences extend to shows within ideological categories as well. Random subsamples of shows within each ideological type revealed surprisingly weak correlations between shows in terms of subject matter.

The most striking differences were found between the issues emphasized by Rush Limbaugh and the content of any other talk radio program studied, particularly other conservative shows. Among other things, Limbaugh is far more likely to emphasize matters pertaining to the free market and to emphasize the value of political efficacy and optimism. These differences, as well as the enormity of Limbaugh's audience relative to that of other hosts, necessitate that Limbaugh be treated separately in this description.

Limbaugh

Originally a host on a local station in Sacramento, California, Rush Limbaugh began his syndicated talk show in New York in 1988. Now heard on more than 650 stations nationwide as well as on shortwave and Armed Services Radio, Limbaugh reaches between fifteen and twenty million listeners per week (Capella, Turow, and Jamieson 1996). As *Talk Daily* (Adams Research 1995) has written, "It is nearly impossible to find an inhabited place in the U.S. where the *Rush Limbaugh Show* cannot be found on the dial." Many people even gather in bars and restaurants across the country to collectively listen in "Rush Rooms." Moreover, Limbaugh is the author of two best-selling books and a monthly newsletter with 170,000 subscribers.

Limbaugh is by far the most popular voice in political talk radio. A *Talk Daily* survey (Adams Research 1995) revealed that nearly 40 percent of all talk radio listeners listen to Limbaugh. Furthermore, among 486 respondents in the 1995 ANES Pilot Survey, 22 percent reported listening to Limbaugh at least occasionally, and 9 percent reported listening at least once a week. Of more importance, 27 percent of *voters* reported listening to Limbaugh, and 15 percent reported listening to Limbaugh at least once a week.[2]

As noted earlier, countless politicians, pundits, and journalists have credited Limbaugh with having considerable power over the contemporary American political landscape. For example, in introducing the featured speaker at a gathering in honor of the seventy-three Republican freshmen of the Congressional class of 1994, former Congressman Vin Weber remarked: "Rush Limbaugh is really as responsible for what has happened [the Republican majority in Congress] as any individual in America. Talk radio, with you in the lead, is what turned the tide" (Kurtz 1996:21). The same freshmen Republicans also named Limbaugh an honorary member of Congress. In a different venue, former education secretary William Bennett declared Limbaugh "the most consequential person in political life at the moment. He is changing the terms of the debate" (Kurtz 1996:21). Even former President Clinton has lamented that "Limbaugh has three hours to say whatever he wants. And I won't have an opportunity to respond" (Devroy and Merida 1994).

Beyond the comments of public officials, a number of media publications have portrayed Limbaugh as a powerful Republican leader, including the *New York Times, National Review,* and *Mother Jones* (Jamieson, Capella, and Turow, 1996).

In terms of the demographics of Limbaugh's audience, Limbaugh listeners are more likely to be white males over the age of fifty than are other conservative talk radio listeners. They also tend to have lower annual incomes, and are less likely to have college degrees.[3] Politically and ideologically, Limbaugh listeners are far more likely to see themselves as conservative (70%) and Republican (61%) than even those who listen to other conservative programs (48% and 45%, respectively).

Limbaugh's mock egotism and bombastic style allude to his role as an entertainer (Limbaugh 1992). However, Limbaugh does not shrink from the part of Republican mouthpiece. "Not only am I a performer, I am also effectively communicating a body of beliefs that strikes terror into the heart of even the most well entrenched liberals, shaking them to their core" (Limbaugh 1994). Indeed, some have called Limbaugh the modern equivalent of the partisan press (Jamieson, Capella, and Turow 1996). As Limbaugh has said, "I think you people can be persuaded. I believe that the most effective way to persuade people is not to wag a finger in their face but to speak to them in a way that makes them think that they reached certain conclusions on their own."[4]

As noted earlier, the topics that Limbaugh chooses to emphasize differ substantially from those of his conservative talk radio brethren or the mainstream press. Capella and his colleagues' content analysis of the Limbaugh program during the primary season of 1996 reveals that Limbaugh appears far less likely than other political talk show hosts to talk about foreign affairs, family issues, education, public ethics, human rights, or crime. On the other hand, Limbaugh appears more inclined to talk about the Clinton administration's job performance, Republican candidates, Congress, and third-party candidates. Furthermore, Limbaugh appears substantially more likely to extol the virtues of the free market and personal efficacy/public optimism than are other talk show hosts or the mainstream media (Capella, Turow, and Jamieson 1996).

To determine the extent to which the messages analyzed by the Annenberg scholars are typical, I scanned summaries of Limbaugh's show, provided on the Internet by John Switzer for the years 1993–95,[5] filtering first for all issues for which there were corresponding questions in the American National Election Studies, and then for politically relevant groups and prominent politicians. The numerical values in parentheses in appendix A indicate the number of *days* during 1993, 1994, and 1995 that those particular issues, groups, and political personalities were mentioned on the show.[6]

In short, this count supports the Annenberg findings. Limbaugh's message appears to focus primarily on the virtues of individual initiative, the free market, and the Republican party, while attacking the media, Ross Perot, anyone associated with the Clinton administration, and groups that Limbaugh perceives as standing in the way of economic freedom. In so doing, Limbaugh devotes considerably less time to other salient issues such as abortion, other moral/cultural issues, and foreign policy. Finally, while Limbaugh routinely makes comments that many would consider as reflecting a certain degree of sexism, he does not appear to espouse overt racism.

I also perused the summaries to gain understanding of Limbaugh's specific message regarding the issues, groups, and public figures on which he focused during that time. Appendix A also provides representative snippets of Limbaugh's message toward government intrusion in the free market, the media, environmentalists, feminists, President Clinton, former Senator Dole, and entrepreneur/former presidential candidate H. Ross Perot. As expected, Limbaugh strongly condemns the federal government, the media, President Clinton, environmentalists, and feminists. Limbaugh's negative portrayal of Independent (or Reform Party) candidate Perot was not surprising, considering Limbaugh's strong association with the Republican Party, and further supports the Annenberg findings regarding Limbaugh's message toward anyone who threatens the "new Republican establishment."

Somewhat unexpected was the finding that Limbaugh expressed great ambivalence toward former Senator Dole during 1993–95. The Annenberg study reflects this ambivalence early in the primary campaign of 1996 (Capella, Turow, and Jamieson 1996). In a separate analysis, Jones (1997) reported a dramatic change in the direction of Limbaugh's coverage of Dole over the course of the 1996 primary season, becoming increasingly supportive as it became apparent that Dole would secure the Republican nomination.

So if Limbaugh attempts to influence his audience while he entertains them, what are the persuasion techniques he employs? Limbaugh's special insert to his *Limbaugh Letter,* entitled "How to Stay Prosperous and Free in the Twenty-first Century," offers a glimpse of the quintessential Limbaugh. A patriotic treatise championing the free market, personal initiative, and conservative public policy, the essay exhibits an array of rhetorical as well as heresthetical propaganda devices. As discussed in chapter 1, I distinguish between *rhetoric,* or the attempt to persuade by modifying what an audience member believes about a given issue, group, or public figure, and *heresthetic,* or the attempt to persuade by framing issues in such a way as to manipulate the salience of particular considerations in memory.

With rhetorical devices, belief structures may be manipulated by making appeals to either reason or emotion (logos or pathos, in Aristotelian terms). Limbaugh's "How to Stay Prosperous and Free in the Twenty-first Century" essay (Limbaugh 1998) contains at least twenty-nine rhetorical appeals to reason, including fourteen personal stories provided to supply "evidence" supporting an argument Limbaugh is trying to make, and fifteen instances of "card stacking"—the arranging of evidence to support an argument (Institute for Propaganda Analysis 1939). The pamphlet also makes extensive use of emotional rhetoric, including seven testimonials from "average, ordinary Americans" designed to make the argument more reliable and trustworthy, and fifteen uses of the "transfer" technique—the attempt to link an idea, group, or person to the reputation of another perhaps unrelated idea group, or person, either by direct name calling or through indirect association.

In terms of heresthetic, Limbaugh strategically and consistently frames his oratory around particular value dimensions at the expense of others. By priming freedom and self-reliance as salient considerations, Limbaugh is able to "butter up" message receivers for the anti-government rhetoric that follows. In the fifty-one paragraphs that constitute "How to Stay Prosperous and Free in the Twenty-first Century" (Limbaugh 1998), Limbaugh makes reference to some core American value such as freedom or self-reliance in forty-one paragraphs (80%). Furthermore, in the thirty-six paragraphs that at least implicitly deal with domestic economic policy (71%), Limbaugh primes the value of freedom in twenty paragraphs and the value of self-reliance in fourteen paragraphs. Sometimes both of these values are primed in the same paragraph; but in all, one of these two related values is primed in twenty-eight out of thirty-six (78%) of the paragraphs concerning domestic economic policy. Moreover, these preferred values are primed in sixteen of the first seventeen paragraphs of the treatise, setting the mood for the specific policy proposals that follow.

By comparison, humanitarian, egalitarian, and communitarian values are scarcely mentioned. Equality, either in name or in spirit, is primed in only five paragraphs, and in one of those paragraphs, the value is portrayed negatively. Three other times, equality is mentioned inside a quotation from one of the founding fathers, which also includes a reference to freedom or self-reliance. Similarly, community (broadly conceived) is primed in none of the paragraphs. Humanitarianism is primed in ten paragraphs, but usually in reference to how people can help others "in the long run" by encouraging them to be self-reliant.

No less important and consistent with the Annenberg findings, Limbaugh's treatise is also replete with efforts to encourage personal and political efficacy among his audience members. This technique demonstrates Limbaugh's fundamental call to action. Although not often urging listeners to make phone calls or to take other specific actions, Limbaugh indirectly urges listeners to participate by trying to make them feel as if they can make a difference, and as if they must.

Such mobilization efforts represent an important facet of persuasion that often takes a back seat to persuasion that attempts to change attitudes. But successfully persuading people to act may be more difficult than just changing their minds—and may have more direct and immediate political consequences—because real activity is engendered, rather than just thoughts.

Summary

In sum, political talk radio has a large, knowledgeable, and active audience that has captured the attention and imagination of journalists, politicians, and pundits. Rush Limbaugh's program—the undisputed leader of the medium—differs significantly in content from other conservative talk radio and may even fill the space in society left vacant by the extinction of the traditional partisan press. Limbaugh, acting as the mouthpiece for the GOP, takes up the conservative cause in its entirety, but spends most of his time discussing matters pertaining to national government spending, the media, feminists, environmentalists, and the Clintons. Limbaugh's message is clear, consistent, entertaining, repeated day after day, and corroborated by a continual stream of callers who agree with Limbaugh. The purpose of his show is "to make him look good," but he readily admits his intent to persuade. While he uses a full range of rhetorical devices to spread his message, Limbaugh consistently frames arguments around preferred value dimensions in an effort to prime listeners to use those values as the bases for their decisions. Furthermore, he tries to make listeners feel as if they have the capacity to make a difference in politics, and therefore he makes abstract, nonspecific urges to his listeners to participate in politics. These characteristics of the Limbaugh message provide essential grounding for the empirical-effects chapters that follow. From these observations, I expect that Limbaugh will persuade audience members to be more conservative on economic matters but not on cultural matters, to engender greater commitment to the Republican party, to prompt listeners to feel more efficacious and thus to encourage participatory behavior, and to do all of this by framing discussion around the core democratic values of freedom and self-reliance, at the expense of equally salient values such as equality, community, and tolerance.

The following chapters attempt to evaluate the degree to which Limbaugh succeeds in his efforts to persuade. Chapter 3 stays in the laboratory in order to monitor the effectiveness of value heresthetic as a propaganda tool relative to rhetoric. Chapter 4 leaves the lab and considers Limbaugh effects as they relate to opinion and vote choice in national general elections. Chapter 5 examines Limbaugh effects in the context of the 2000 Republican primary battle between George W. Bush—Limbaugh's preferred candidate—and John McCain, whom Limbaugh strongly opposed. Chapter 6 addresses persuasion from a mobilization perspective, analyzing the degree to which listeners become emboldened or stifled in response to Limbaugh. Chapter 7 examines political information and misinformation, to see if there is any correspondence between talk radio listening and public understanding. Chapter 8 serves to wrap up the entire book by reviewing the major theoretical and applied political questions addressed in the preceding chapters, summarizing the results, and discussing the possible implications of these findings. A primary methodological goal to be addressed in each chapter concerns distinguishing persuasion effects from correlations that are attributable to a self-selected talk radio audience. Various methods are employed to disentangle meaningful effects from coincidental associations.

3 Toward a Value Heresthetic Model of Political Persuasion

David C. Barker, Kathleen Knight, and Christopher Jan Carman

Before we can properly assess the degree to which talk radio may persuade listeners to think and behave in predictable ways, it is necessary to understand the ways in which such persuasion may occur. Given that democratic politics revolves not around coercion of the public but rather around the struggle to persuade others that one choice is better than another, understanding the dynamics of this struggle is fundamental to any meaningful understanding of modern politics.

The previous chapters introduced *heresthetic* as a theoretical construct distinct from traditional rhetoric, reviewed its treatment in the literature (primarily from a framing/priming point of view), and established its centrality to the Limbaugh message. This chapter empirically examines the power of heresthetic as a persuasion tool, relative to rhetoric, within a controlled experimental environment. Experimental designs provide the best opportunity for researchers to assess causality, as opposed to simple association, between variables of interest—in this case, message exposure and policy preference. By holding the message sender, subject, context, and medium constant, and by randomizing audience exposure to different messages, we were able to evaluate more precisely the relative persuasive impact of heresthetic and rhetoric, at least within our laboratory setting.

To review, both heresthetic and rhetoric are tools at a propagandist's disposal when trying to induce some belief, choice, or behavior. Rhetoric involves attempting to persuade an audience by providing the audience with new information—even if that information is designed to appeal primarily to emotion rather than reason. The audience member considers something new (e.g., "Now, we make our toothpaste with an additional drop of retsin" or "Have a Coke and a smile") and, it is hoped (from the propagandist's perspective), updates his or her opinion or level of motivation based on that new information. Heresthetic, on the other hand, is more about strategic choices. Using heresthetic does not involve providing new information. Rather, it involves framing messages in such a way as to prime particular considerations (which already exist in the audience's consciousness) to the front of the audience's collective head. It sets the audience members' cognitive agenda, so to speak. It manipulates which considerations will be considered salient to the question at hand.

A number of theoretical reasons lead us to hypothesize that heresthetic outperforms traditional rhetoric as a persuasion determinant. First, as detailed in chapter 1, the use of heresthetic may neutralize the role of audience sophistication in the persuasion equation. Although people vary widely in their degree of political knowledge, ability, and interest—making belief change far more difficult to induce when addressing a sophisticated audience—heresthetic capitalizes on that which is commonly stored in memory, making persuasion at least as likely when audiences are sophisticated as when they are unsophisticated, because the message does not challenge the audience members' beliefs.

Second, the reputation or credibility of the source may not be as relevant to the persuasion situation when heresthetic is being employed as it would be under rhetorical conditions, because the message sender is not, on the surface, trying to provide new information or otherwise change what the audience believes in terms of factual knowledge. For example, audience members may reject information provided by a disreputable source, but they are less likely to respond negatively to a message that encourages them to think in terms of some cherished value. In the latter case, the audience member does not have to decide whether to believe the information being given by the source—thus rendering source credibility irrelevant. Put another way, by neutralizing the roles of audience sophistication and source credibility, respectively, heresthetic may discourage cognitive counterargument by audience members.

Furthermore, heresthetic may persuade by *activating* core values. A large body of research has now been accumulated to suggest that core values, or matters of principle (meaning "conceptions of the desirable, not just something to be desired" Kluckhohn 1951:395) motivate and guide political judgments to a much greater extent than do rational cost-benefit calculations of the desired (Rokeach 1973; Lane 1973; McClosky and Zaller 1984; Hochschild 1995; Feldman 1988; Stoker 1992; Hurwitz and Peffley 1987; Hurwitz, Peffley, and Seligson 1993). In other words, citizens do not usually make political decisions by determining "what's in it for me," but rather "what is right; what is wrong" from a normative perspective. Moreover, Americans may be particularly inclined to make value judgments because Americans, it is often argued, are relatively united in their commitment to a common, small number of principles that have guided the political development of the republic (e.g., Tocqueville [1848] 1945; McClosky and Zaller 1984; Kinder 1998).

Perhaps foremost among this commonly held set of values stands *individualism* (Feldman 1988), the belief that each person should be self-reliant and free to pursue his or her interests, accepting full responsibility for the consequences of those pursuits. Coined in the aftermath of the French Revolution, individualism was associated with incivility and social chaos by Europeans (Burke [1790] 1910). However, in the United States, individualism has always been revered as a moral virtue (McClosky and Zaller 1984). When Zaller and Feldman (1992) asked respondents to provide explanations of their attitudes toward the role and responsibility of the federal government, the researchers found that those who oppose federal spending and expansion of services were much more likely to moralize about individualistic themes in their responses. Indeed, Americans tend to place blame for social decay and unemployment not on the whole of society but on the impoverished themselves (Sniderman and Brody 1977; Feldman 1983; Gurin et al. 1969; Kluegel and Smith 1986).

But while the utility of values as agents of attitude formation and change has been clearly demonstrated, a relative paucity of scholarly attention has been given to how latent value considerations become activated, or why some value considerations become activated at the expense of others (Kinder 1998). *We argue, quite simply, that values become activated via the influence of social networks, both electronic and interpersonal.* In other words, propagandists can manipulate which (if any) values are activated during an audience member's cognition toward a particular political object.

At this point, it is important to note that we do not argue that momentary exposure to propaganda actually instills values. Rather, we contend merely that such exposure to persuasive messages may prime which, if any, value considerations come to the "top of the head" when making political choices. This point will be elaborated on formally in the next section.

We further contend that the activation of particular values instead of others has significant consequences regarding individual opinions toward specific issues, and that the competition between different values for prominence within the American ethos partially explains the considerable policy ambivalence found among the American public (Zaller 1992; Alvarez 1998). As Tetlock (1986) explained, Americans are stricken with what might be called "value pluralism." In other words, most of us, to some extent, simultaneously cherish different sets of political values that suggest opposing policy outcomes. For example, although freedom and equality are not necessarily irreconcilable, either principle, if taken to its logical absolute, thwarts the policy goals of the other: How can we guarantee all citizens an equal opportunity to a quality public education without stripping away some of the individual liberty of local school boards and taxpayers, and vice versa? The struggle to expand civil rights in the United States for Americans traditionally denied liberty is also the story of government restricting the "liberty" of some Americans in the name of a moral commitment to equality.

Empirical analysis has borne out the idea that exposing audiences to different sets of value considerations engenders different policy preferences. For example, Katz and Hass (1988), in examining the ambivalence of white Americans toward minorities, found that whites held more positive attitudes toward blacks after the values of humanitarianism and egalitarianism had been primed than when the Protestant work ethic had been primed.

In light of such findings, we expect that audiences exposed to messages adulating individualism are more inclined to oppose federal spending toward the poor than those exposed to messages void of explicit value content but replete with rhetorical appeals. As expressed earlier, we expect that heresthetic trumps rhetoric as a persuasion tool.

The following section provides a more detailed, formal explanation of this value heresthetic model and the predictions it generates regarding persuasion, relative to the traditional rhetorical model that has guided much social-psychological persuasion research to date.

A Model of Value Heresthetic, Rhetoric, and Persuasion Through Talk Radio

This section attempts to further explicate the distinction between rhetoric and heresthetic and develops a theory of the relative persuasive power of the two general techniques by offering a formal model of the persuasion process within the talk radio context. To clearly understand heresthetic and rhetoric as distinct persuasion strategies, particularly within the context of how Limbaugh may influence opinion formation and change, consider the following hypothetical scenario. The Democrats have just proposed $20 billion in new federal spending to expand the Food Stamp program. Conservatives oppose the measure. Rush Limbaugh has instructed his call screener to give preference to callers who want to voice their opinions regarding the Democrats' new spending proposal. Limbaugh's goals are twofold: (1) to provide entertainment, in order to protect and expand his listenership and (2) to encourage listeners to oppose the new spending proposal. Focusing on the persuasive goal, it can be said that Limbaugh's preferences are also transitive. That is, Limbaugh wants listeners to become more opposed to the spending after exposure to his message, but he prefers the status quo to a "boomerang effect"—*increased* listener support for the spending bill after Limbaugh exposure.

Figure 3.1 illustrates a hypothetical Limbaugh audience member *i*'s opinion regarding spending on food stamps at time-point *t*, prior to Limbaugh exposure. To simplify, we designate $\hat{y}_t$ to represent this pre-exposure opinion. On this scale, "1" equals firm opposition to the new spending bill, while "-1" equals unwavering support for the bill. As figure 3.1 shows, $\hat{y}_t = 0$. In other words, prior to Limbaugh exposure, audience member *i* is undecided; he or she is no more likely to support new spending than oppose it.

Figure 3.2 depicts *i*'s attitudes toward two values, individualism and egalitarianism, which are potential considerations pertaining to $\hat{y}_t$. We designate x_2 to represent *i*'s point of view toward individualism. A score of "1," located on the right side of the graph, equals ardent support for the individualistic principles of personal freedom and self-reliance, while a score of "-1" equals perfect opposition to such principles.

Correspondingly, we designate x_1 to represent *i*'s attitude toward egalitarianism. For egalitarianism, a score of "-1," located on the left side of the graph, equals spirited support for the principle of equality. We display support for egalitarianism on the left side of the graph be-

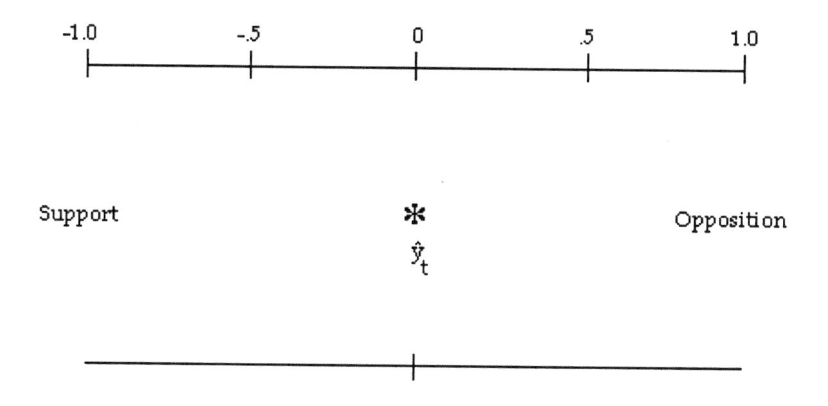

Figure 3.1 Attitude: Spending Bill Prior to Limbaugh Exposure

cause egalitarianism may be thought of as a value associated with the political "left," while individualism, at least in the economic realm, is often associated with the political "right." It is these commitment scores that a propagandist may try to manipulate through the use of new information, or rhetoric. Figure 3.2 reveals that at *t*, *i* strongly endorses both individualism and egalitarianism. In notational form,

$$\hat{y}_t = a + x_1 + x_2$$

or

$$0 = 0 + 1 - 1$$

where *a* symbolizes *i*'s latent attitude toward spending on food stamps, shown to be 0 in this case because, as noted earlier, *i* is undecided, having not yet carefully thought about the issue. For the sake of simplicity, we have depicted these models as deterministic, or without an error term. Of course, in reality, there are innumerable variables that influence individual policy preferences, which would be captured in a (considerable) error term. Our example is not intended to provide a comprehensive account of individual decision making, but rather to enhance understanding of how value heresthetic and rhetoric may operate in the business of directing political choice through exposure to talk radio.

Figure 3.2 also displays the likelihood that each value consideration will be a salient determinant of $\hat{y}_t$, as illustrated by the density of lines

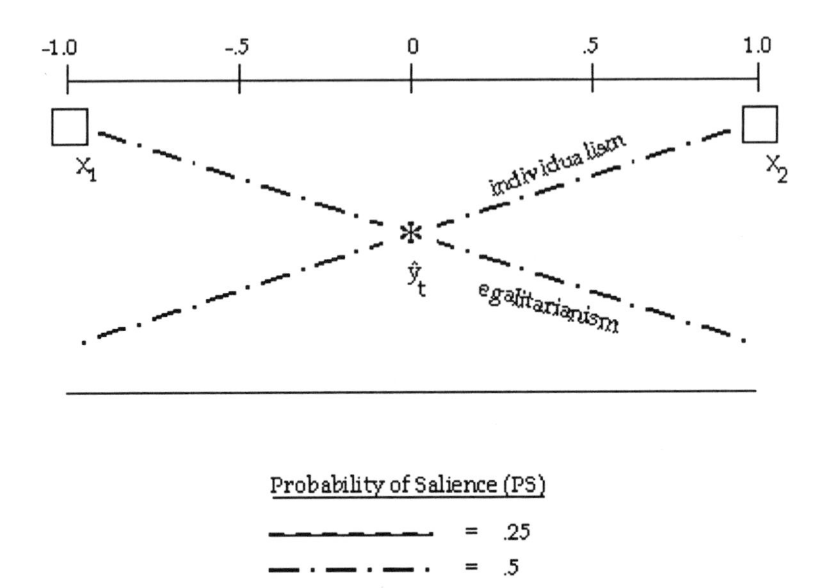

Figure 3.2 Attitude: Spending, Individualism, and Egalitarianism Prior to Limbaugh Exposure

x_1 and x_2. This is distinct from how committed the audience member is to these principles. For those familiar with statistics or econometrics, these salience scores may be thought of as beta weights (b). It is these beta weights that heresthetical appeals attempt to manipulate. In this example, both individualism and egalitarianism have beta weights of .5, meaning each consideration has an equal chance of being highly salient to *i*'s judgment toward increased spending on food stamps. Including consideration salience in our model adjusts our equation of $\hat{y}_t$ to resemble that of a standard regression model:

$\hat{y}_t = a + b\ 1x1 + b\ 2x2,$

or

$\hat{y}_t = 0 + .5(1.0) + .5(-1)$

Given Limbaugh's preferences, what strategy might he employ to persuade *i*? Limbaugh's screening device informs him that the first caller, a moderate from Omaha, supports the new spending, and will argue that income disparity in the United States is greater than in any other Western democracy. Thus, the Nebraskan caller, upon achieving airtime, will frame the issue in egalitarian terms, priming egalitarianism as a salient consideration of $\hat{y}$. In determining how to proceed, Limbaugh is faced with a dilemma in terms of his persuasion strategy. Recall that the target of the persuasion attempt is not necessarily the caller but rather the listening audience. On one hand, Limbaugh could employ rhetoric, arguing with the caller about the level of income disparity in the United States. In so doing, he would concede to the caller that this issue should be thought of in egalitarian terms. On the other hand, Limbaugh could pursue a strategy involving heresthetic. Again, and at the risk of sounding redundant, in contrast to rhetoric, which essentially manipulates beliefs regarding considerations that are already salient, heresthetic attempts to manipulate the salience of considerations already believed. Therefore, if Limbaugh chooses heresthetic as a persuasion strategy, he will attempt to redefine the issue in terms of freedom and self-reliance.

Figures 3.3 and 3.4 depict $\hat{y}_{t+1}$, after exposure of *i* to the verbal exchange between Limbaugh and the caller. Both represent scenarios in which Limbaugh's message had resonated with *i*. Figure 3.3 illustrates a successful application of rhetoric as a persuasion tool, while figure 3.4 depicts a successful application of heresthetic. Recall that the caller had initially framed the debate in egalitarian terms, priming egalitarianism to the top of *i*'s cognitive processing. As such, a .75 probability now exists that egalitarianism will be salient to *i*'s judgment regarding spending on food stamps.

In the rhetoric scenario portrayed in figure 3.3, Limbaugh had conceded the dimension of relevant considerations to the caller, and had attempted to induce movement in $x1$—the listener-in-question's level of egalitarian sentiment. As figure 3.3 shows, rhetoric works to some extent in convincing *i* that inequality, measured as income disparity, may have positive qualities if (as he argues) it is a function of relative work effort. So at $t + 1$, after exposure to Limbaugh, *i* has become less enthusiastic in his or her egalitarianism (at least in the short term). However, *i*'s baseline level of egalitarianism is so strong (−1) that even a 25 percent decrease in commitment still leaves *i* strongly supportive of the

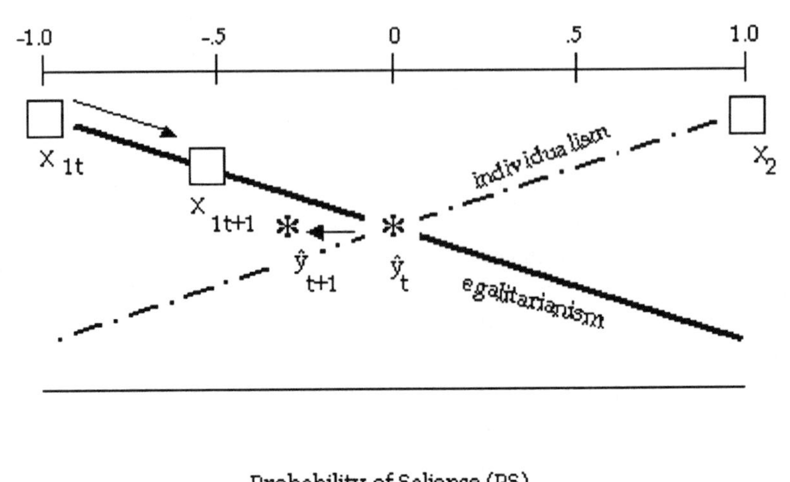

Figure 3.3 Attitudes After Exposure to Rhetoric

principle, and because Limbaugh has conceded that the battle will be fought on egalitarian grounds, *i* is now more likely than before to think in egalitarian terms (.75 to .5) when forming a decision on this issue. Consequently, *i* actually became *more* likely to support the new food stamp proposal at $t + 1$ than he or she had been at t. So the mathematical equation for $t + 1$ becomes:

$$\hat{y}_{t+1} = 0 + (-.75).5$$

or

$$\hat{y}_{t+1} = -.375$$

In other words, after exposure to the exchange between Limbaugh and the caller in which Limbaugh employed rhetoric but conceded the framing of the debate to the caller, the listener, *i*, is 12.5 percent more like-

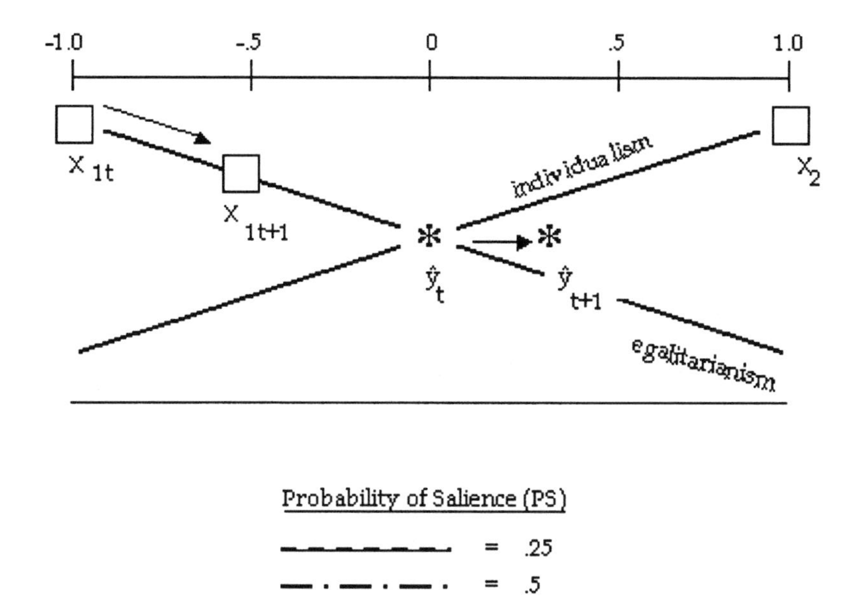

Figure 3.4 Attitudes After Value Heresthetic

ly to support the new spending proposal than before exposure to the verbal exchange. Limbaugh's efforts boomeranged.

Such an outcome is obviously undesirable in terms of Limbaugh's preferences. Figure 3.4 illustrates the second scenario, after Limbaugh has pursued a strategy involving the use of heresthetic. After replying briefly to the caller's egalitarian argument, Limbaugh had successfully shifted the dimension of the debate to considerations of individualism. Speaking in abstract terms, Limbaugh argued that individuals, not government, should assume responsibility for their own financial well-being. Therefore, Limbaugh capitalized upon *i*'s strong and established commitment to individualism as a principle, and encouraged the listener to think along such lines. As a consequence, the listener *i* became 25 percent more likely to consider the principle of individualism when making his or her judgment toward the policy. So our equation of probability of support for new food stamp spending after exposure changes again:

$\hat{y}_{t+1} = a + b\ 1x1 + b\ 2x2$

or

$\hat{y}_{t+1} = 0 + (.75)-.5 + (.75)1.0$

or

$\hat{y}_{t+1} = .375$

Thus by simply shifting the dimension of the discussion to prime individualism as a relevant consideration, Limbaugh increased the likelihood that *i* would oppose the new spending bill at $t + 1$ by nearly 19 percent, relative to $\hat{y}_t$, and by nearly 38 percent relative to $\hat{y}_{t+1}$ when rhetoric had been the only persuasion strategy employed. Most important, the use of heresthetic led $\hat{y}_{t+1}$ to cross the threshold from likely support of the bill to likely opposition.

To summarize, this section has attempted to provide an example of how a propagandist may use rhetoric and heresthetic to influence the dynamics of opinion formation for specific targeted audience members. We have sought to show how heresthetic may be an effective and preferred method of propaganda in some situations. By taking advantage of commonly shared values that, if primed, suggest particular policy preferences, a propagandist may effectively induce the opinion he or she seeks from audience members without inviting cognitive counterargument on the part of audience members. By reducing the risk of counterargument, the propagandist reduces the likelihood that an audience member will become *less* supportive of the propagandist's position after exposure to the message. Hence, the potential costs associated with pursuing heresthetic as a persuasion strategy, both in terms of effort and risk, will often pale in comparison to the potential costs of employing rhetoric, making the use of heresthetic a rational persuasion strategy in many instances. The following sections provide an empirical examination of this process, using controlled experiments to see the differential impact of value heresthetic and non-value-based rhetoric on audience members' opinions toward federal spending on the poor.

Experimental Analysis

This section uses controlled experiments to empirically examine the relative utility of value heresthetic, which deliberately and strategically frames arguments around chosen core principles, and non-value-based rhetoric, which attempts to persuade with informational appeals to either reason or emotion. Subjects were randomly exposed to one of three different edited messages, two of which involved Rush Limbaugh making persuasive appeals. One Limbaugh recording contained only non-value-based rhetoric. The second Limbaugh recording contained rhetoric as well, but also contained arguments designed to prime the values of self-reliance and economic freedom. A third cell of subjects, the control group, received a message providing political information but no persuasive appeals of any kind.

Recruitment and Subject Profile

Three weeks before the first experimental session, we began recruiting University of Houston undergraduates[1] to act as experimental subjects. Interested students filled out contact information sheets and (in most cases) pretest surveys[2] containing questions regarding political partisanship, ideology, issue positions, exposure to and affect toward Rush Limbaugh, and standard socioeconomic/demographic variables.[3] Participation was encouraged by the promise of food and drink, as well as a fifty-dollar prize to be awarded to one out of every ten participants. To ward off testing effects caused by subject awareness (Campbell and Stanley 1963), subjects were told before the sessions began that the goal of the research was to determine the extent to which distractions affect students' ability to process and recall specific bits of political information.

Table 3.1 describes the demographic, socioeconomic, partisan, and ideological makeup of the subjects. With regard to basic demographics, the sample is disproportionately young (as would be expected in a sample of undergraduates), female, and African American. With respect to socioeconomic status, the sample is drawn from reasonably well-to-do families, again reflecting what one would expect from a sample of college students, even at a public urban university. Although most of the subjects' fathers did not graduate from college, the mean family income is in the thirty- to fifty-thousand-dollar range. In terms of sophistication regarding public affairs, the sample appears reasonably well informed, tending to see the economy as having improved, the deficit as having declined, and the average tax burden as having stayed about the same since the early 1990s. Moreover, eighty-one of ninety-one subjects knew that a two-thirds majority vote is needed for Congress to override a presidential veto, and eighty subjects knew that the Republicans were

Table 3.1 Subject Profile

Variable	Mean	*N*	Standard Deviation	Pretest/ Posttest
Age	22	55	6.78	Pre
Fundamentalist	1.89	55	.63	Pre
White	.50	55	.50	Pre
Black	.32	55	.63	Pre
Hispanic	.16	55	.37	Pre
Female	.60	90	.4	Post
Family Income	3.30	55	1.17	Pre
Warmth toward Clinton	60.30	55	28.70	Pre
Warmth toward Limbaugh	23.80	55	25.52	Pre
Party ID	2.80	55	1.76	Pre
Ideology—self-ID	2.80	55	1.53	Pre
Spending	3.40	55	1.31	Pre
Abortion	3.50	55	2.08	Pre
Knowledge: Economy	2.10	90	.76	Post
Knowledge: Deficit	3.20	90	1.20	Post
Knowledge: Tax	2.80	90	.85	Post

Note: The variables are coded as follows:

Fundamentalist: 1 = Bible is literal word of God, 2 = Bible is inspired word of God, 3 = Bible is just a book;
White: 1 = white;
Black: 1 = black;
Hispanic: 1 = hispanic;
Female: 1 = female;
Family Income: 1 = below $12K, 2 = $12K–$30K, 3 = $30K–$50K, 4 = $50K–$80K, 5 = over $80K;
Clinton: 0 = intense hostility, 100 = intense warmth;
Limbaugh: 0 = intense hostility, 100 = intense warmth;
Party ID: 0 = strong Democrat, 6 = strong Republican;
Ideology: 0 = intense liberalism, 6 = intense conservatism;
Spending: 1 = preference for many more services and much more federal govt. spending, 7 = preference for many fewer services and much less govt. spending;
Abortion: 1 = pro-choice in all circumstances, 7 = pro-life in all circumstances;
Economy: 1 = much better since early ’90s, 5 = much worse since early ’90s;
Deficit: 1 = much larger since early ’90s, 5 = much smaller since early ’90s;
Tax: 1 = much larger for those making less than $100K since early ’90s, 5 = much smaller for those making less than $100K since early ’90s.

the majority party in the House of Representatives. Such knowledge was expected from a sample of students drawn from American government courses. Unfortunately, there was not enough variance in the sample to undertake meaningful analyses of differential effects according to audience sophistication. However, the higher-than-average level of political sophistication on the part of the subjects may have served to attenuate the extent to which persuasive appeals could affect subject attitudes (Zaller 1992).

Furthermore, the subjects were, on the whole, slightly liberal and Democratic. They displayed liberal attitudes toward government spending and abortion, and were remarkably supportive of President Clinton. The subjects' mean feeling thermometer score (0–100) for Clinton, measuring personal affect rather than job performance assessment, was 60, even in the face of scandalous allegations of sexual misconduct that broke in the news only days before these measures were taken. Furthermore, the pretest subjects were strikingly opposed to Rush Limbaugh, exhibiting a mean feeling thermometer score of only 23 toward the conservative radio personality.

While the subjects can hardly be said to constitute a representative sample of the public at large, neither are they entirely typical "college sophomores in the lab" (Sears 1986). Of more importance, these statistics paint a picture of a sophisticated audience somewhat hostile toward the product that Limbaugh is trying to sell—conservatism. Because many scholars have demonstrated that persuasion is much less likely to occur when the audience is not at least somewhat inclined to agree with the message sender (e.g., Bennett 1980), and when the audience does not perceive the message sender to be a credible and trustworthy source (e.g., Hovland, Janis, and Kelley 1953), Limbaugh's message was forced to swim upstream in its attempt to persuade this audience.

Specific Procedures

Each experimental session took approximately twenty minutes. After subjects read and signed detailed consent forms, they received personal cassette players with headphones and were randomly assigned to one of two rooms for the experimental session. Proctors randomly gave subjects one of three cassette tapes (A, B, or C). Proctors in the room were "blind" as to which letter represented which experimental condition, thus eliminating the possibility of proctors matching sociodemographic characteristics of subjects with particular tape stimuli.

At this point, each subject recorded the version of the tape to which he or she was listening on the cover sheet of his or her posttest questionnaire, but did not look at the questionnaire. The proctor, reading from a script, then instructed the subjects to begin listening to their cassette tapes, using the players and headphones provided. Individual headphones provided the intimacy with the message necessary to ensure not only that individual subjects would be ignorant of the message to which other subjects were listening, but also to better simulate the environment in which most people listen to talk radio (while driving by themselves). Given the particular age cohort of the subjects (late teens to early twenties), the pseudo-intimacy of headphones was a natural condition, because many members of this age cohort have spent years listening to music on such devices. To encourage comfort and ward off boredom, participants were provided with food and soft drinks, as well as pencils and several sheets of paper. The opportunity to distract oneself from the message by eating and/or doodling provided our design with another obstacle for persuasion to overcome. Our goal throughout was to create an environment as hostile to persuasion as possible, to ward off the possibility that any persuasion found was an artificial condition of the testing environment.

After listening to one of three 11-minute recordings, subjects completed posttest questionnaires containing (1) several political knowledge items designed to convince subjects that they were participating in a learning study, (2) several questions measuring sociodemographic characteristics and partisanship, (3) questions measuring value preferences, and (4) questions measuring federal spending preferences and ideological attachments.

The Stimuli

The control stimulus consisted of eleven minutes of political geography lessons from *Don't Know Much About Geography,* by Kenneth Davis (1992). Specifically, students heard statistics about American geographical knowledge compared to that of other industrialized nations, followed by a short glossary of terms, differentiating between a "republic," a "state," a "nation," a "principality," and so on. The two other cells of experimental subjects listened to excerpts from Davis as well, but they were also exposed to excerpts from Rush Limbaugh's second book, *See, I Told You So* (1993). Holding the messenger constant between the rhetoric and value heresthetic cells was necessary in order to control for messenger effects between the two cells.

In both manipulations of the Limbaugh exposure, the specific issue that Limbaugh discussed was also held constant. In both manipulations, Limbaugh ultimately argues that liberal programs (i.e., federal government spending) designed to ameliorate social ills are wrongheaded. By keeping the issue dimension uniform, we control for any priming influence that the specific issues themselves might have on persuasion.

Although the messenger and the issue are constant across both experimental manipulations, the specific arguments made in each stimulus differ considerably. One message consists entirely of rhetorical appeals to emotion and reason, attempting to provide listeners with new information to be used in evaluating the worth of government spending on the poor. This message is devoid of explicit appeals to core values. We classify this message as the *rhetoric* stimulus. Appendix B contains extended excerpts from the rhetoric stimulus.

Although the *value heresthetic* stimulus also contained an appreciable dose of rhetoric, in this message, the rhetoric was preceded by an extended discourse on the sanctity of individual liberty, economic freedom, and self-reliance. This device attempted to alter the dimension by which subjects based their opinions, prompting subjects to concentrate on cherished core values. Specifically, Limbaugh talked about "removing the shackles" of government intervention, freeing individuals to be the "best they can be" by discovering their potential through rugged self-reliance. Appendix C contains extended excerpts from the value heresthetic stimulus.

Specific Hypotheses

H1: When faced with a decision between competing value considerations, subjects exposed to the value heresthetic stimulus are more likely to prefer self-reliance to humanitarianism than subjects exposed to either the control stimulus or the rhetoric stimulus.

H2: When faced with a decision between competing value considerations, subjects exposed to the value heresthetic stimulus are more likely to prefer economic freedom to equality of opportunity than subjects exposed to either the control stimulus or the rhetoric stimulus.

H3: Subjects exposed to the value heresthetic stimulus are more likely to oppose federal spending to assist the poor than subjects who are exposed to either the control stimulus or the rhetoric stimulus.

Selection Bias?

In large samples, random assignment of the experimental stimuli ensures that all experimental groups are roughly equal in terms of the distribution of previously held attitudes (Cook and Campbell 1966). However, with only ninety-one cases, even a random sample may produce substantial differences between groups. These differences between groups can have a profound effect on the influence of the stimuli. For example, if conservatives had been disproportionately exposed to the value heresthetic stimulus, observed differences in attitudes toward government spending between those exposed to the value heresthetic stimulus and the other groups would likely be inflated. Simply stated, even with random assignment of stimuli, it may be necessary to control for audience demographics in experiments with small samples.

Table 3.2 Mean Differences Between Experimental Groups in General Ideological Disposition

Stimulus	Variables*			
	Ideology	Cons-Libs	FT Limbaugh	Spending
Control	3.03	-2.88	27.50	3.73
Rhetoric	3.13	-2.83	25.07	3.67
Value Priming	2.87	-12.22	20.00	3.53
Total	3.01	-5.98	23.83	3.64
N	55	90	55	55

None of the differences displayed in this table is statistically significant.
*Pretest measures, except Cons-Libs, which was only measured in the posttest questionnaire.
Note: The variables are coded as follows:
Ideology: 0 = intense liberalism, 6 = intense conservatism;
Cons-Libs: 0 = no difference in warmth toward conservatives versus liberals, −100 = intense positivity toward liberals and extreme negativity toward conservatives, 100 = intense positivity toward conservatives and extreme negativity toward liberals;
FT Limbaugh: 0 = intense hostility toward Limbaugh, 100 = intense warmth toward Limbaugh;
Spending: 1 = preference for many more services and much more federal govt. spending, 7 = preference for many fewer services and much less govt. spending.

As stated earlier, this sample as a whole exhibited slightly left-of-center preferences and partisanship. But were differences present between groups (value heresthetic, rhetoric, or control) in terms of the attitudes carried into the experiment? Table 3.2 shows the differences in means between the different groups with regard to general ideological disposition and attitudes. All but one of these attitudes were measured in a pretest questionnaire to which only about half of all respondents were exposed. As the table shows, differences between groups are not statistically significant. Statistically significant differences were not expected in a sample of this size, but one substantive relationship stands out. The group exposed to the value heresthetic stimulus appears predisposed to dislike Rush Limbaugh to an even greater extent than the other groups. The mean feeling thermometer score for Limbaugh among those receiving the value heresthetic stimulus is 20, compared to 25.1 for those receiving the rhetoric stimulus, and 27.5 for those receiving the control stimulus. These differences in means suggest that while the entire sample leaned to the left, the value heresthetic group leaned at a decidedly sharper angle, stacking the deck against the persuasive utility of Limbaugh's value heresthetic message.

Results

The first step in evaluating the value heresthetic hypothesis involves testing the degree to which being exposed to a political message that primes one set of values over others results in a measured tendency to judge that value as more important than other competing values. More precisely, do those exposed to Limbaugh's adulation of self-reliance and the free market exhibit markedly stronger tendencies to see those values as more important than equality of opportunity and humanitarianism?

Preference for Self-Reliance is measured by posttest responses to the question: "Is it more important to be helpful and cooperate with others, or to encourage self-reliance?" Responses are dichotomous. "Zero" indicates preference for humanitarianism; "one" indicates preference for encouraging self-reliance. Figure 3.5 displays the difference in means for this variable according to which experimental manipulation the subject received. Among the thirty subjects exposed to the control stimulus, the mean score was .18, indicating a clear preference among the control subjects for helping others. Among the thirty subjects exposed to the Limbaugh rhetoric stimulus, the mean score was .37, indicating some

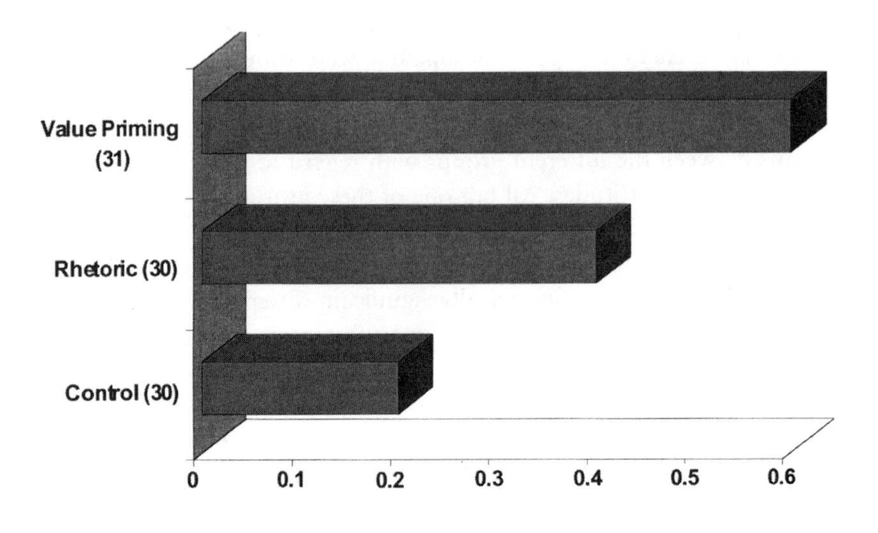

Figure 3.5 Mean Differences in Value Preference
(0 < Helping Others; 1 < Self-Reliance)

movement toward a preference for self-reliance, but still showing a mean preference for helping others. By contrast, the thirty-one students who received the Limbaugh value heresthetic stimulus displayed a mean score of .60, a higher mean preference for self-reliance than subjects in the other two cells demonstrated. As a one-way ANOVA (analysis of variance) illustrates, these differences are statistically significant ($p<.01$), and the stimulus explains 33 percent of the variance in value preference among the subjects. This simple test suggests that attempts to prime the value of self-reliance, rather than humanitarianism, indeed produces a value preference for self-reliance.

Preference for Economic Freedom is measured by posttest responses to the question: "Is it more important for the federal government, through its policies, to encourage economic freedom, or that everyone has a fair chance?" "One" indicates preference for equality of opportunity; "two" indicates preference for economic freedom. Figure 3.6 displays mean differences for this variable according to the experimental stimulus. The mean scores for the control and rhetoric stimuli were virtually indistinguishable, hovering at .31 and .30, respectively. In contrast, the mean score of those receiving the value heresthetic stimulus was .47, reflecting a stronger tendency to prefer economic freedom over equality of opportunity. These differences are also statistically signifi-

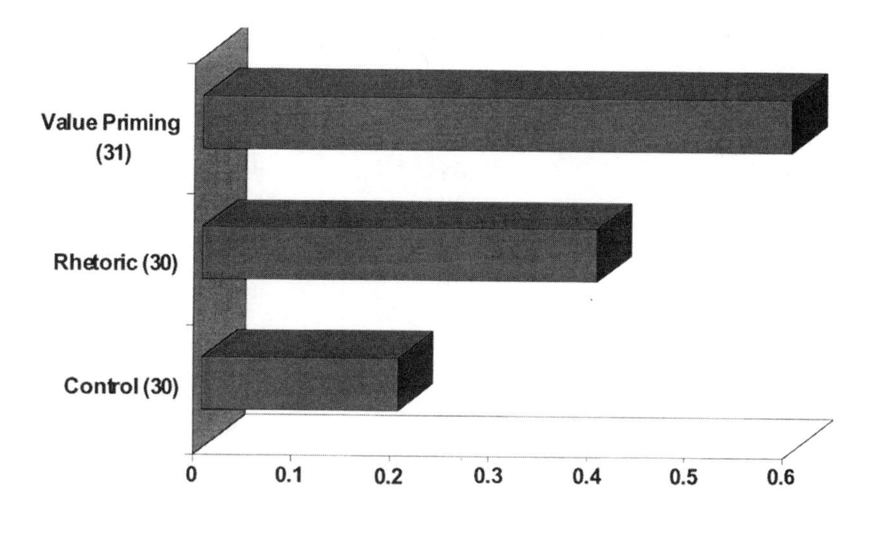

Figure 3.6 Mean Differences in Value Preference
(0 < Equality; 1 < Freedom)

cant ($p<.05$), and the stimulus explained 18 percent of the variance in preference for economic freedom.

Table 3.3 displays results from an *ordinary least squares* (OLS) regression analysis of democratic value preference, combining freedom and self-reliance to form a measure of *individualism.* The dependent variable is a two-item index, ranging from zero (preference for both humanitarianism and equality of opportunity) to two (preference for both self-reliance and economic freedom). The independent variables of interest, (1) exposure to value heresthetic, and (2) exposure to rhetoric, are measured as dichotomies. In the case of value heresthetic, "one" equals exposure to the value heresthetic stimulus; "zero" equals exposure to either the control or rhetoric stimuli. In the case of rhetoric, "one" equals exposure to the rhetoric stimulus; "zero" equals exposure to either the control or value heresthetic stimuli. As would be the case with regression with dummy variables, the control group is the reference category. Control variables include gender, exposure to the pretest survey, attitude toward spending on health care, self-identified ideology, attitude toward government spending in general, and the difference in warmth toward liberals and conservatives.

As table 3.3 illustrates, exposure to the *rhetoric* stimulus produces a .29 unit (15%) increase in support for self-reliance and freedom. This re-

Table 3.3 Multivariate OLS Regression Analysis of Democratic Value Preference

Persuasion Technique	*B*	Std. Error	Beta
Value Priming	.64***	.19	.40
Rhetoric	.29*	.20	.18
N	90		
Adj. R^2	.16***		

To facilitate presentation, the following control variables are not displayed in the table, but are present in the analysis: gender, general spending preference, exposure to the pretest survey, ideology, opposition to spending on health care, difference in affect toward conservatives and liberals, and the constant term.
Note: The dependent variable is trichotomous (0—2)—"0" signifies preference for equality over economic freedom and philanthropy over self-reliance; "1" indicates ambivalence; "2" equals preference for economic freedom over equality and self-reliance over philanthropy.

* $p < .10$, one-tailed test;
*** $p < .001$, one-tailed test.

lationship is not statistically significant, but it points to some persuasive utility. However, exposure to the value heresthetic stimulus produces a .64 unit (32%) increase in propensity to value individualism more than equality and humanitarianism, holding everything else constant. This relationship produces a coefficient-to-standard-error ratio (*t*) of 3.38, indicating a strong, statistically significant relationship that would differ from zero in 999 out of 1000 independently drawn samples. Therefore, it appears that Limbaugh succeeds in inducing value preferences, even in the face of considerable preconceived opposition to his message. He succeeds to some extent even when he does not explicitly discuss the values of self-reliance and economic freedom. But his impact more than doubles when he generously decorates his message with references to the importance of these values. Nevertheless, does such value heresthetic translate into more-conservative sentiment regarding federal social spending?

Table 3.4 reports results of a multiple logistic regression analysis of opinion toward spending to help the poor. Again, the independent variables of interest are the persuasion techniques to which subjects were exposed. As was the case with the previous analysis of value preference, we controlled for gender, general and health care spending preferences,

Table 3.4 Multiple Logistic Regression of Attitudes Toward Spending

Persuasion Technique	*B*	Std. Error	Exp. (*b*)	Impact[a]	Impact[b]	Impact[c]
Heresthetic	1.68**	.84	5.34	39%	34%	19%
Rhetoric	.03	.89		0	0	0
N	91					
Nagelkerke R^2	.46***					

To facilitate presentation, the following control variables are not displayed in the table, but are present in the analysis: gender, general spending preference, exposure to the pretest survey, ideology, opposition to spending on health care, difference in affect toward conservatives and liberals, and the constant term.
Note: The dependent variable is dichotomous (0—1)—"0" signifies preference for more or the same amount of federal government spending to assist the poor; "1" signifies preference for less federal government spending to assist the poor.

[a] Estimated increase in the probability of opposing spending on the poor when other variables leave the chance of opposition at 25%.
[b] Estimated increase in the probability of opposing spending on the poor when other variables leave the chance of opposition at 50%.
[c] Estimated increase in the probability of opposing spending on the poor when other variables leave the chance of opposition at 75%.

** $p < .05$, two-tailed test;
*** $p < .001$, two-tailed test.

exposure to the pretest questionnaire, and self-identified ideology. Converting the logistic regression coefficients into estimates of increases in the odds ratio demonstrates that exposure to the rhetoric stimulus engenders no change in the odds that an individual will oppose spending on the poor. Conversely, individuals exposed to the value heresthetic stimulus were *more than five times* as likely to oppose spending on the poor, holding everything else constant.

Determining the influence of the stimuli in terms of *probability* change is more tricky because the probability of opposing spending on the poor differs according to the prior probability level of opposition. As the final three columns in table 3.4 show, if the prior probability of an individual opposing spending on the poor was only .25, exposure to

Figure 3.7 A Structural Equation of Exposure to Different Propaganda Techniques, Value Preference, and Opinion

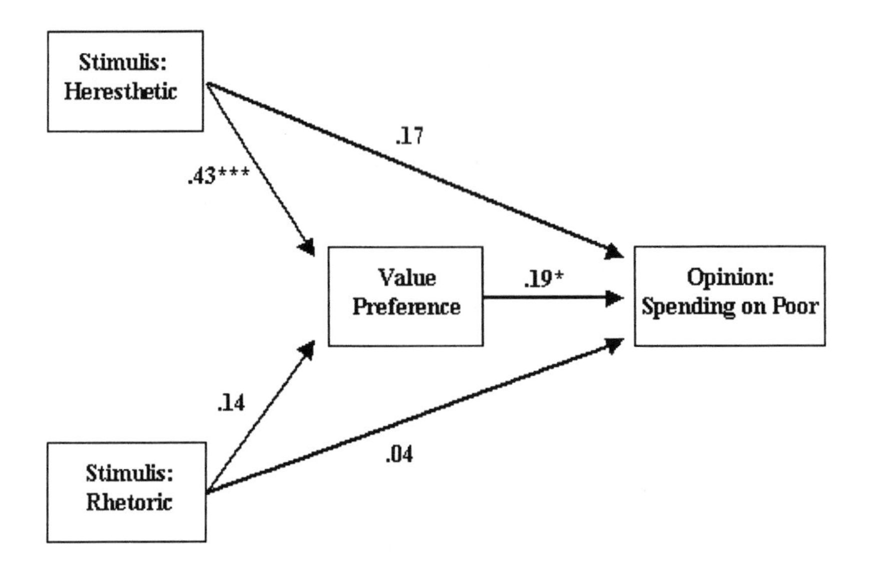

To preserve clarity, we have not drawn the following control variables in this depiction, although they are included in the empirical model: gender, race, exposure to the pretest, pilot participation, and opinion on health care spending.

Note: Comparative Fit Index > .99; coefficients are standardized.

* $p < .05$, one-tailed test
*** $p < .001$, one-tailed test

the value heresthetic stimulus had the greatest impact in terms of real change, making them 39 percent more likely to be opposed. If a person was "on the cusp," at 50 percent probability of opposition, then exposure to value heresthetic increased the probability of opposition by 34 percent. Finally, for those who were already inclined to oppose spending on the poor ($P = .75$), the stimulus increased the probability of opposition by 19 percent.[4]

Figure 3.7 displays the standardized coefficients and test statistics for a structural equation,[5] depicting the causal path of message exposure, value preference, and opinion toward spending on the poor. This equation enables us to simultaneously observe the direct and indirect effects of exposure to value heresthetic and provides us with an estimate of the degree to which our theory "fits" the data. As expected, exposure to the value heresthetic stimulus is strongly associated with a preference for individualism over egalitarian or humanitarian concerns, which translates into a meaningful increase in the likelihood of opposition to spending on the poor, statistically significant in a one-tailed test—the appropriate test, given our expectations from the regression analyses. This structural model provides substantial support for our causal model, namely, that value heresthetic induces value preferences, which in turn guide policy preferences. The nearly perfect comparative fit index indicates that no alternative specification of the causal model produces a more accurate reflection of the true relationships found in the covariance matrix than this one.

Discussion

This chapter examined the extent to which framing issues in such a way as to encourage message receivers to think in one value dimension versus another can encourage different policy preferences. We hypothesized not only that such value heresthetic is an effective means of persuasion, but also that under the right conditions, value heresthetical appeals may engender much more persuasion than do standard rhetorical appeals. Based on an experimental research design to maximize the internal validity of our analysis, and sampling techniques and procedures designed to minimize the likelihood that value heresthetic would succeed persuasively, our findings provide support for our hypotheses. By priming the values of economic freedom and self-reliance, as opposed to other equally cherished values such as humanitarianism and equality of opportunity, Rush Limbaugh was able to increase the likelihood that subjects would oppose federal spending to assist the disadvantaged more than fivefold. This apparent persuasion occurred even though those who were exposed to the Limbaugh value heresthetic stimulus reported, on average, strong and disproportionate hostility toward the talk show host, considerable negativity toward conservatives, displayed liberal tendencies in terms of general attitudes toward the role of government, and endured distractions throughout the experiment. Moreover, the sample was composed almost entirely of subjects with higher-than-average levels of political sophistication—people who are least susceptible to persuasion according to Zaller's RAS (reception acceptance sampling) model of opinion.

The effectiveness of value heresthetic as a political persuasion technique is strengthened by the finding that Limbaugh had little influence when trying to persuade by using traditional rhetorical appeals to either reason or emotion. These findings support Schattschneider's prescient claim that "the definition of alternatives is the supreme instrument of power. He who determines what politics is about runs the country" (Schattschneider 1960:68).

However, a couple of caveats are in order. First, the setting under which we tested our hypotheses, one could argue, was quite artificial. Even though we tried to create an environment that was as natural as possible, any controlled experiment, with only ninety-one volunteer subjects from a single U.S. city, will not produce results that carry a great deal of inferential power.

Second, our ability to manipulate the Limbaugh message in order to distinguish between techniques was less than optimal. To create a setting that more closely resembled reality, we chose to use actual excerpts from Limbaugh, deleting some phrases, rather than creating arguments ourselves and then attributing them to Limbaugh. Although this decision certainly paid dividends in terms of our ability to draw conclusions about the persuasive power of Limbaugh (at least parts of his message), relying on the actual message made the contradistinction between the heresthetic and rhetoric stimuli a little more "fuzzy" than we would have liked. As noted earlier in the chapter, both of the experimental stimuli contained elements of heresthetic and rhetoric. The primary distinction between the two involved the presence of higher-order *values* as the considerations being primed in the value heresthetic stimulus. Future research should work to distinguish between value-based rhetoric, value-based heresthetic, non-value-based rhetoric, and non-value-based heresthetic. Furthermore, the effort should be made to pit competing values against one another, to see the relative power of framing social issues around the values of egalitarianism and humanitarianism versus individualism.

Third, our research design did not enable us to examine the stability of the apparent opinion inducement. One month after collecting our original data, attempts to recontact subjects found that most of the subjects had changed residences (returning, presumably, to their permanent addresses for summer vacation). Measures of the temporal persistence of persuasion via value heresthetic would greatly enhance our understanding of the utility of value heresthetic as a persuasion determinant. However, one should not conclude that short-term persuasion is irrelevant. First of all, short-term persuasion is one mechanism that may lead to lasting attitude adjustment. As the on-line model of opinion formation and adjustment has demonstrated repeatedly (e.g., Lodge and McGraw 1995), individuals appear to frequently adjust their attitudes, even if only slightly, in response to newly encountered stimuli. As soon as they update their attitude toward a particular political object, however, people forget the reasons for the update. Over time, people may develop quite strong attitudes without necessarily being able to articulate the reasons underlying their preferences. Repeated exposure to a stimulus, such as the priming of a particular value contained within a daily talk radio show, may (over time) breed lasting attitudes.

But studying short-term attitude change is not merely a means to an end. Even if the attitudes brought on by value heresthetic last no longer than the time it takes to provide a survey response, understanding their origin is essential to contemporary democratic theory. In today's political climate, nary a political nor a policy-related decision is made without great consideration being given to how that decision will play out with the electorate (Morris 1998). Such considerations become manifest not so much by reading constituent mail or monitoring phone calls, but rather through the painstaking analysis of carefully crafted and sampled surveys of public opinion. Thus the survey response has taken on such a significant role in the determination of campaign strategy and policy pursuit that the etiology of that survey response has become a compelling puzzle in its own right.

The following chapters move away from examining *how* persuasion through talk radio may occur, to the more-general question of *whether*, outside the controlled environment of the lab, persuasion via talk radio occurs at all.

4 Talk Radio, Public Opinion, and Vote Choice: The "Limbaugh Effect," 1994–96

David C. Barker and Kathleen Knight

The previous chapters introduced talk radio as a medium indicative of the "new media," examined message content, and tested hypotheses about the relative persuasibility of different oratorical strategies that message senders may employ in trying to win over an audience. We found that value heresthetic, or the framing of a debate in such a way so as to prime considerations based on core democratic values, may be a particularly useful means of achieving persuasion, particularly when an audience is sophisticated or predisposed to distrust the message source. These findings were made possible through content analysis and carefully controlled experiments manipulating Limbaugh messages and recording differences in attitudes in response to those messages. However, the laboratory setting of those experiments imposed an artificiality to the analysis that calls into question the degree to which talk radio listening results in any kind of attitudinal or behavioral changes in the "real world." This chapter attempts to rectify that problem to some extent, by examining how attitudes, preferences, and vote choices correspond to exposure to the Limbaugh message, and how such exposure is associated with changes in judgment over time. We thus move away from hypothetical choices made in classrooms over pizza in front of video

---

Parts of this chapter are reprinted with permission from "Talk Radio Turns the Tide? Political Talk Radio and Public Opinion" (Barker and Knight 2000).

cameras to real preferences of real people surveyed nationally with sophisticated sampling techniques. In this chapter, we focus on persuasion in terms of attitude change, as expressed through policy preferences, candidate evaluations, and vote choices—the central elements of republican government.

Thus in the following analysis we employ data from the American National Election Studies to consider what difference it really makes that millions of Americans listen to Limbaugh every day. Could it be that regular Limbaugh listeners make choices in large part based on exposure to the political deliberation encountered on the Limbaugh show?

One simple way to conclude that a message is persuasive is to determine that individuals who have heard the message are in greater agreement with its content than are those who have not heard it. The classic problem with this method is selection bias—individuals already in agreement with the message content may choose to expose themselves to commentators who share that point of view. This chapter uses a variety of techniques in the attempt to overcome this problem. We take a first cut at overcoming this problem by looking at the relationship between Limbaugh listening and opinions about a range of issues, groups, and political actors, controlling for various influences. Then we distinguish between the topics that Limbaugh discusses frequently and those he addresses only on an occasional basis. If substantive relationships exist between listening and opinions, regardless of whether Limbaugh discusses the target issues with regularity, we may infer that Limbaugh likely has little independent influence—the observed relationships exist because listeners are conservative and Republican, even if our error-laden measures of these concepts do not fully capture the extent of their influence. However, if consistent relationships emerge between listening and opinion regarding matters that Limbaugh stresses, but not for issues to which he gives less attention, we will be more inclined to conclude that Limbaugh's independent influence is real.

We take a second stab at overcoming selection bias by examining attitudes toward a figure for whom the Limbaugh message runs independent of predictable conservative sentiment and/or the Republican "party line." For this portion of our analysis, we focus on respondent warmth toward former presidential candidate H. Ross Perot. The bivariate relationship between conservatism and warmth toward Perot is positive, but Limbaugh's treatment of Perot's bid for the presidency was relentlessly negative (Jamieson, Capella, and Turow 1996). Therefore we hypothesize that regular listening to Limbaugh should have a significant negative effect on evaluations of Perot.

Of course, even if we are able to demonstrate that listeners seem to be persuaded counter to their partisan or ideological predisposition while nonlisteners do not, this does not rule out the possibility of self-selection altogether. We therefore reexamine each of the multivariate relationships between listening and opinion with a model that substitutes an instrumental variable of Limbaugh listening for the actual measure. The instrument was created by constructing a model of Limbaugh listening composed entirely of items that do not correlate theoretically or statistically with conservatism. By doing this, we are better able to parse the shared variance of Limbaugh listening and conservatism, making selection bias far less troublesome.

Finally, we analyze Limbaugh influence by taking into account the necessary temporal component of causality. We take advantage of the panel component of the data to investigate whether and to what extent regular Limbaugh listening leads to attitude change consistent with Limbaugh's messages over time.

Methodological Issues

In nonelection years, the American National Election Study Board (ANES) often conducts a pilot study aimed primarily at evaluating new instrumentation. These studies are usually "piggybacked" on the regular election studies by selecting a subsample for reinterview. This allows for experiments on question wording and creates a small panel of reinterviewees. The crucial variables for the analyses we undertake here are found in the 1995 ANES non-election-year pilot study. The 1995 pilot study included a Limbaugh feeling thermometer and several substantially more-detailed media consumption questions, including whether respondents listen to the Limbaugh radio show and other political talk radio shows.

The design of the pilot studies poses some interesting analytic challenges. The Limbaugh feeling thermometer is available for both 1993 and 1995, but these panel components are attached to opposite random half-samples of the 1994 ANES congressional election study. This means that we cannot, for example, calculate the correlation between Limbaugh thermometers in 1993 and 1995—the detailed media questions are available in 1995 only. Although the 1995 panel is small (n = 486), it is a representative sample of the American public, interviewed after the 1994 election, again six to eight months later, and twice in 1996—before and after the presidential election.

Traugott and colleagues (1996) examined a two-wave panel survey conducted by the Times Mirror Center for the People, the Press, and Politics in 1994. However, they focused on subsequent attention to news rather than attitudes or attitude change. These researchers looked separately at domestic and foreign policy issues. They reasoned that because political talk radio, and Limbaugh in particular, focused almost exclusively on domestic policy issues in 1994, listening to talk radio or to Limbaugh should predict subsequent attention to domestic news, but not to foreign news. In the analysis undertaken below, we engage in a similar exercise, but one focused directly on opinion. We hypothesize that listening to Limbaugh will significantly influence the direction of opinion about issues, groups, and people Limbaugh discusses frequently, and that listening to Limbaugh will not be a significant predictor of the direction of opinion about issues, groups, and people that Limbaugh mentions relatively infrequently. As stated earlier, if a linear relationship emerges between the magnitude of the "Limbaugh effect" and frequency of mention, then the hypothesis that Limbaugh has persuasive influence will have received support.

Limbaugh and Public Opinion–Cross-sectional Evidence

We have analyzed the partial relationship between regular Limbaugh listening and opinion measured in the 1995 ANES pilot, for which the literature and our understanding of the Limbaugh message allowed us to predict a direction of association (e.g., we did not analyze attitudes toward the AARP because we have no clear indication of either the conservative position or the Limbaugh position regarding this group). Altogether, we analyzed opinions toward twenty items: Vice President Gore, President Clinton, Senator Dole, Speaker Gingrich, General Powell, Senator Gramm, Governor Wilson, Hillary Clinton, President Clinton's degree of liberalism/conservatism, the news media, environmentalists, religious groups like the Christian Coalition, big business, government spending, the government's role in guaranteeing jobs, preference for environmental protection versus job growth, welfare policy, the deficit, defense spending, and taxes.

The question is if, controlling for other factors, knowing whether and how much a respondent listens to Limbaugh helps to predict the direction and strength of opinion. As stated earlier, we expect that Limbaugh's influence on listeners extends primarily to those issues that Limbaugh emphasizes. First, we ran a series of regressions predicting each of the items listed above by Limbaugh listening,[1] controlling for age, race, geographic region, mainstream media attendance, income, education, being a homemaker, trust in government, gender, fundamentalist religiosity, party ID, self-identified ideology, and affect toward Limbaugh himself.[2] In order to further identify the unique influence of Limbaugh, separate from that of other conservative messages that may reiterate dominant themes, we also control for exposure to other political talk radio shows. In an attempt to measure exposure to conservative talk radio other than Limbaugh, we also multiply exposure to other political talk radio by respondent ideology. Similarly, in order to control for exposure to other conservative media, we also include an interaction term that multiplies respondent ideology by exposure to mainstream media—newspapers, news magazines, and TV news (local and national). Thus this model takes great care to ward off the influence of self-selection and spuriousness.

Second, we placed each topic of discussion (the dependent variables in the previous regression equations) in a category ranging from 1—"rarely discussed" (discussed on the Limbaugh show on fewer than 150 days during 1993–95, which would have amounted to less than once a week, on average)—to 5—"constantly discussed" (discussed on more than 600 days during 1993–95, amounting to discussion on virtually every day). The number of days in 1993–95 that a topic was discussed was determined by scanning the transcripts provided on the Internet by John Switzer (see chapter 2 for more details). Categories divide evenly by increments of 150 days, which represent average increases of one day per week. The topics that are classified as "rarely discussed" (0–150 days) include Phil Gramm and conservative Christians. Topics classified as "infrequently discussed" (151–300 days) include Gingrich, big business, defense spending preference, opposition to the deficit at all cost, support for Powell, and support for Wilson. Topics classified as "somewhat frequently discussed" (301–450 days) include support for Dole, jobs versus environmental protection, government activity regarding job creation/insurance. "Frequently discussed" (451–600 days) topics include Al Gore, Hillary Clinton, environmentalists, welfare, and whether Bill Clinton is a moderate or a liberal. "Constantly discussed" (more than 600 days) topics include federal domestic spending, Clinton's job performance, Clinton's personal character, and the news media (as measured by an index of feeling thermometer scores for the major news anchors: Shaw, Brokaw, Jennings, and Rather).

Third, we extracted the standardized partial regression coefficients of the Limbaugh listening predictor from each model. We then created a new data set, with topic serving as the unit of analysis. With the extracted standardized regression coefficients of the relationships between Limbaugh listening and opinions toward the various topics mentioned above, we created a dependent variable called "degree of association." In this new data set, with topic as the unit of analysis, we then regressed the degree of association variable on the measure of Limbaugh salience (average number of days mentioned per week). If Limbaugh has independent influence, the standardized partial coefficients should be larger for opinions regarding topics he discussed regularly than for topics he discusses marginally or rarely. We therefore expected to find a strong linear relationship between topic salience on the Limbaugh show and degree of association between listening and conservative opinion toward that topic.

Figure 4.1 displays the scatter plot of this relationship. This scatter plot displays a clear linear trend. An increase in the average frequency of mention for an issue of roughly one day per week is associated with a .79 standard deviation change in the degree of association between

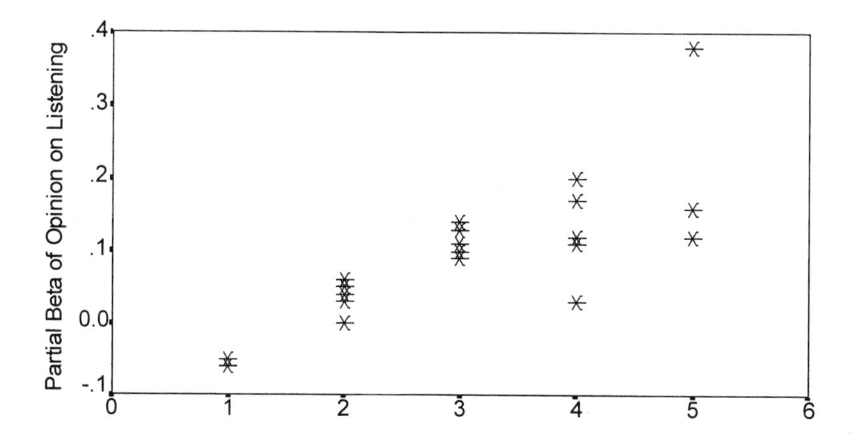

Figure 4.1 Scatter Plot of the Frequency with Which an Issue Is Discussed by Limbaugh, as It Accounts for the Strength of the Relationship Between Limbaugh Listening and Opinion—1995. (0 = Helping Others; 1 = Self-Reliance)

listening and opinion. Stated more dramatically, movement from being rarely mentioned to being constantly mentioned is related to a 3.2 standard deviation change in degree of association, enough to produce a change of more than 70 percentile ranks.

As figure 4.1 reveals, the magnitude of this slope is significantly accounted for by a single outlier—the extraordinarily strong relationship between frequency of mention and association between listening and attitudes toward the news anchors. Still, even when controlling for the presence of this outlier, movement from being rarely discussed to being constantly discussed is associated with a 2.5 standard deviation change in degree of association. Of course, Limbaugh appears to be more effective when he is spouting hostility rather than positivity toward a subject. But even when controlling for the direction of Limbaugh's message toward a topic (positive/negative), movement from being rarely mentioned to being constantly mentioned still corresponds to a 2 standard deviation change in degree of association. The degree to which Limbaugh may encourage positive affect for some ideas, groups, or individuals is a topic to which we return in a later section.

These results lend support to the hypothesis that the correlation between Limbaugh listening and conservative opinion is not entirely a function of selection bias. An even more convincing case can be made by employing the two-stage least-squares technique, which we describe in the following section.

Two-Stage Least-Squares Analysis

One common method of dealing with models involving reciprocal causality, such as the one we have here, is to perform *two-stage* least-squares analysis. In the first stage of this technique, the researcher creates a proxy, or instrumental variable to take the place of some observed independent variable in a regression equation. This instrumental variable is created by treating the observed independent variable as endogenous, regressing the observed variable on a series of predictors that are related to the observed variable but theoretically and empirically unrelated to the dependent variable or other control variables in the original model from which the observed independent variable was drawn (in this case various measures of political and ideological conservatism). The predicted scores of this model are saved, and those saved scores become the instrumental variable that, in the second stage, replaces the original variable in the equation. Because the resulting in-

Table 4.1 Constructing an Instrument of Limbaugh Listening

Predictors	b	std. error
Constant	−.22	.08
Miles driven weekly	.0003	.0001**
Political interest	.05	.02*
Partisan	.07	.03*
Political discussion	.11	.03***
N	489	
Adjusted R^2	.12	

Source: 1994–95 American National Election Study

* $p < .05$
** $p < .01$
*** $p < .001$

strument is a reflection of the original independent variable, absent the portion that may be caused by that dependent variable, researchers create instruments in the hope of gaining a clearer picture of the real causal direction between two correlated variables.

In creating an instrumental variable of Limbaugh listening, the variables used to predict Limbaugh listening in the construction of the instrument are unrelated to ideological conservatism and include: (1) the number of miles that a respondent drives in a week, (2) the frequency with which a respondent talks about politics, (3) the degree to which a respondent is a partisan (either Democrat or Republican), and (4) the degree to which a respondent is interested in politics and public affairs. Table 4.1 displays the results from the first stage of this analysis—the construction of the instrument. As the table shows, each of these predictors is statistically related to Limbaugh listening, even though they are all independent of ideological concerns. Taken together, these predictors explain 12 percent of the variance in Limbaugh listening, a reasonable proportion, given that surely a substantial amount of the variance must be explained by party identification and ideology.

Tables 4.2 and 4.3 depict the second stage of the analysis, where the predicted scores of Limbaugh listening from the first stage are substituted as an instrument of Limbaugh listening within multiple regression

Table 4.2 2SLS Estimates of Limbaugh Impact – Frequently Discussed Topics

Predictors	Dependent Variables							
	Clinton	Perot	Media anchors	Crime bill	Health	Govt. role insurance	Feminists	Environ-mentalists
	b (std.e.)	*b* (std.e.)	*b* (std.e.)	Logit (std.e.)	*b* (std.e.)	*b* (std.e.)	*b* (std.e.)	*b* (std.e.)
Constant	91.75	29.24	78.15	3.37	.61	1.92	105.82	93.97
	(5.17)***	(6.36)***	(7.18)***	(.78)***	(.43)	(.35)***	(5.71)***	(5.04)***
Biblical literalism	–2.48	.88	1.34	–.24	–.11	.08	–4.37	–1.38
	(2.08)	(2.56)	(2.73)	(.27)	(.17)	(.14)	(2.30)	(2.03)
Income	.08	.20	–.11	.04	.04	.007	-.25	.29
	(.17)	(.21)	(.25)	(.02)	(.01)*	(.01)	(.19)	(.17)
Conservatism	–2.57	1.63	–1.97	–.05	.35	.35	–4.75	–4.43
	(.94)**	(1.15)	(1.33)	(.13)	(.08)***	(.06)***	(1.03)***	(.91)***
Male	–2.97	4.82	–3.51	–.49	-.21	.25	–3.88	–3.67
	(1.93)	(2.38)*	(2.63)	(.26)	(.16)	(.13)	(2.13)	(1.88)
South	–1.59	.23	1.98	–.15	.14	.10	3.59	2.26
	(2.04)	(2.50)	(2.56)	(.27)	(.17)	(.14)	(2.45)	(1.98)
Trust in govt.	4.56	3.18	–2.61	.43	.09	.02	2.72	.86
	(1.23)***	(1.51)*	(1.68)	(.18)*	(.10)	(.08)	(1.36)*	(1.20)
White	–6.68	1.46	7.26	–.56	.28	–.07	–8.26	–4.36
	(3.06)*	(3.76)	(4.18)	(.50)	(.25)	(.21)	(3.38)*	(2.99)
Republicanism	–5.31	–.51	–2.14	-.32	.20	.12	–2.14	-1.52
	(.52)***	(.64)	(.72)**	(.07)	(.04)***	(.002)***	(.04)***	(.50)**
Warmth toward	–.09	.17	.09	–.01	.005	–.003	–.06	.02
Limbaugh	(.04)*	(.05)***	(.05)	(.00)*	(.003)	(.003)	(.04)	(.04)
Limbaugh	–11.22	–18.22	–12.21	–1.51	1.02	.86	–11.60	–10.94
listening	(4.26)**	(5.24)***	(6.21)	(.59)*	(.35)**	(.30)	(4.70)*	(4.15)**
instrument								
N	486	486	203	486	486	486	486	486
Adj. R^2	.40	.07	.15	.19	.22	1.38	.25	.16

Source: 1994–95 American National Election Study
* $p < .05$; ** $p < .01$ *** $p < .001$

equations in order to get an assessment of the relationship between the portion of Limbaugh listening that is uncontaminated by party identification/ideological sentiment and various political preferences and attitudes. Looking at table 4.2 first, this table displays the relationship between the instrument of Limbaugh listening and opinion regarding matters that Limbaugh emphasizes on his show, as determined by our own content analysis of the shows between 1993 and 1995, as well as the detailed analysis conducted by Capella, Turow, and Jamieson (1996) at the Annenberg School of Communication at the University of Pennsylvania (see chapter 2 for more details). We controlled for all of the same items listed in the analysis described in the previous section, but ultimately dropped control items from the final equations that failed to add any meaningful predictive utility to any of the models. The resulting equations predict opinion regarding the following topics: the government's role in providing services and spending (measured on a seven-point scale; 7 = preference for much less spending and many fewer services), national health insurance (seven-point scale: 7 = strong opposition), President Clinton (one-hundred-point feeling thermometer), news anchors (an index of feeling thermometer scores toward Peter Jennings, Dan Rather, Tom Brokaw, and Bernard Shaw), the 1994 Clinton Crime Bill (dichotomous),[3] environmentalists (one-hundred-point feeling thermometer), the women's movement (one-hundred-point feeling thermometer), and Ross Perot (one-hundred-point feeling thermometer).[4]

As can be easily observed in table 4.2, the persuasive ability of Limbaugh appears to have passed a very conservative test—that is, even when all of the partisan dispositions and latent ideological sentiment is removed from the measure of Limbaugh listening, listening still explains opinion in a substantial and consistent way regarding matters that Limbaugh frequently discusses on his show.[5]

One relationship deserves particular attention. Public evaluation of Reform Party founder and former presidential candidate Ross Perot makes for an ideal case study of Limbaugh influence, because Limbaugh's message toward Perot during this time period did not necessarily match the dominant conservative view. Given that Perot is politically independent, we do not expect prior partisan affiliation to filter public opinion toward the billionaire. Empirically, warmth toward Perot is not related to Republican partisanship ($r = .03$). However, our examination of the Limbaugh message revealed that Limbaugh leveled consistent criticism toward the Texas billionaire in 1993–95, reaching its peak during the NAFTA struggle in late 1993. We therefore predict a negative rela-

Table 4.3 2SLS Estimates of Limbaugh Impact – Infrequently Discussed Topics

Predictors	Dependent Variables							
	Abortion	Death Penalty	Family values	Gingrich	Gramm	Preferential hiring	No nuclear proliferation	School prayer
	b (std.e.)	*b* (std.e.)	*b* (std.e.)	Logit (std.e.)	*b* (std.e.)	*b* (std.e.)	*b* (std.e.)	*b* (std.e.)
Constant	.94	3.49	2.81	–.55	13.18	1.94	.02	2.58
	(.23)***	(.31)***	(.22)***	(4.62)	(6.08)*	(.20)***	(.05)	(.30)***
Biblical literalism	.78	.09	–.42	1.80	.46	.07 (.08)	.02	.55
	(.09)***	(.13)	(.09)***	(1.86)	(2.28)		(.05)	(.12)***
Income	–.02	–.03	.004	.10	.26	.02	.006	–.005
	(.008)*)	(.01)*	(.007)	(.15)	(.19)	(.007)*	(.004)	(.01)
Conservatism	.22	–.17	–.24	3.01	2.47	.13	.001	.07
	(.04)***	(.06)	(.04)***	(.84)***	(1.02)*	(.04)***	(.02)	(.05)
Male	.08	–.26	–.04	4.34	3.60	–.25	.10	–.14
	(.09)	(.12)	(.08)	(1.73)*	(2.07)	(.08)*	(.04)*	(.11)
South	–.11	–.18	.06	4.79	3.96	.003	.06	.13
	(.09)	(.12)	(.08)	(1.82)**	(2.12)	(.08)	(.04)	(.12)
Trust in govt.	.03	–.03	.08	3.14	3.27	–.05	–.04	.04
	(.06)	(.07)	(.05)	(1.10)**	(1.41)*	(.05)	(.03)	(.07)
White	–.03	–.36	.07	2.23	–.21	.91	.20	-.38
	(.14)	(.18)	(.13)	(2.73)	(3.79)	(.12)***	(.07)**	(.17)*
Republicanism	–.03	–.03	–.02	2.02	1.41	.008	–.01	.003
	(.02)	(.03)	(.02)	(.46)***	(.58)*	(.02)	(.01)	(.03)
Warmth toward Limbaugh	.004	–.003	.000009	.37	.30	-.0004	.0007	.004
	(.002)	(.002)	(.002)	(.03)***	(.04)***	(.001)	(.0008)	(.002)
Limbaugh listening instrument	.02	.40	.11	2.91	–1.61	-.003	.11	–.34
	(.19)	(.26)	(.18)	(3.80)	(4.55)	(.17)	(.09)	(.24)
n	486	486	486	486	486	486	486	486
Adj. R^2	.91	.07	.18	.42	.17	.17	.04	.09

Source: 1994–95 American National Election Study
* $p < .05$; ** $p < .01$ *** $p < .001$

tionship between regular Limbaugh listening and warmth toward Perot. Hence if Limbaugh listening is highly associated with negative affect toward Perot, as would be predicted from Limbaugh's message, then we will have gone some distance toward overcoming selection bias (at least that which is based on preexisting ideology or party identification).

The empirical results support our expectations. Greater Limbaugh listening is associated with less warmth toward the Texas billionaire. Even though Perot's independence does not rule out the possibility that "Perot-bashers" could have learned of Limbaugh's treatment of the Texan and, as a result, started tuning in to Limbaugh, this brand of selective exposure seems less plausible than the standard filtering that people engage in because of prior partisan and ideological dispositions.

By way of comparison, table 4.3 presents the independent relationship between listening to Limbaugh (the instrument) and holding conservative opinions regarding matters that Limbaugh rarely discusses: school prayer, the prevention of nuclear proliferation, preferential hiring according to race or gender, "family values," abortion, the death penalty, and Phil Gramm. Intuitively speaking, we do not expect Limbaugh to guide opinion without trying. If significant relationships emerge between the listening instrument and opinion toward matters that Limbaugh virtually ignores, then those relationships would likely be a spurious function of some unforeseen and uncontrolled-for variable that corresponds to both listening and opinion on those matters.

The findings indicate that, as expected, Limbaugh listeners appear no more likely to hold conservative opinions regarding topics that Limbaugh rarely discusses than do Republicans and conservatives who tune out Limbaugh. If anything, as in the case of school prayer, Limbaugh listeners appear more liberal than their fellow Republican cohorts.

In sum, the coefficients and corresponding tests of significance for this instrumental analysis corroborate the findings that were obtained in the previous section, when raw measures of Limbaugh listening were included in the equations.[6] We have done our best to control for the widely ranging "other factors" that predict attitudes about political issues, groups, and political actors. We have found that we can increase the accuracy of our prediction of public attitudes by observing whether an individual listens to Rush Limbaugh, but only when Limbaugh has focused clearly and explicitly on that issue, group, or person in the majority of his broadcasts. The consistent nonrelationship between listening and issues Limbaugh does not emphasize lends further persuasive support to the hypothesis that Limbaugh is doing more than preaching to the converted. Limbaugh also appears to be much more effective at stirring up opposition than he is at mobilizing support. Perhaps this is because he spends much more time attacking ideas, groups, and individuals than he spends defending other ones. Of course, the notion that negative speech is more effective than positive speech has strong empirical grounding (e.g., Cobb and Kuklinski 1997).

We have thus included two sets of cross-sectional analyses that point to a Limbaugh effect. Instrumental variables help to combat the nefarious consequences of selection bias in quasi-experimental research, but may be criticized on the grounds that instrumental variables are more error laden than raw measures, producing some questions regarding the reliability of the measure. The fact that strong independent associations exist between Limbaugh listening and a variety of opinions regarding topics frequently treated on the Limbaugh show, regardless of whether the raw or the instrumental measure of Limbaugh listening is used, add to the robustness of our findings.

However, even though the cross-sectional data do travel some distance toward demonstrating a "Limbaugh effect," such a determination cannot be conclusively drawn from cross-sectional associations of any kind. While our use of controls (on both the left-hand and right-hand sides of the equations) mitigate the likelihood that observed relationships are entirely a spurious function of self-selection by listeners, we have not presented any evidence that can rule out that possibility. Because cross-sectional data are restricted to exploring associations at a single time point, our analysis to this point can only infer that listening preceded and led to conservative sentiment. A much more effective means of untangling such causal webs involves studying change over time. How do opinions change over time in response to regular exposure to Limbaugh? The following section addresses this question, taking advantage of the panel component of the data to examine the extent to which regular doses of Limbaugh actually induced listeners to change their minds during 1994–96.

Limbaugh and Opinion Change–Panel Evidence

In 1996, the American National Election Study (ANES) reinterviewed 389 respondents who had previously been interviewed in 1994. This panel of respondents offers the opportunity to examine the degree to which listening to Limbaugh in 1995 was associated with actual

Table 4.4 Panel Analysis of Public Opinion on Limbaugh Listening

Independent variables	Opinion toward Pres. Clinton—'96		Opinion toward Bob Dole—'96		Opinion toward gov't spending—'96	
	b	std. error	*b*	std. error	*b*	std. error
Limbaugh '95	.16	.04**	3.36	1.42**	.07	.04***
Opinion-'94	.73	.04	.46	.05***	.77	.04***
Constant	-.07	.04**	25.54	2.63***	-.02	.03
Adj. R^2	.59		.24		.60	
N	361		389		389	

Source: American National Election Study, 1994–95 panel

* $p < .10$, one-tailed test
** $p < .05$, one-tailed test
*** $p < .01$, one-tailed test

changes in opinion by respondents between 1994 and 1996. By including a lagged measure of respondent opinion as a predictor variable in a model predicting opinion in 1996, we are able to control for all the factors that were associated with listening both in 1994 and 1996, leaving the remaining variance in opinion in 1996 to represent *changes* in opinion. In this section, we look at the relationships between Limbaugh listening and changes in opinion toward two of Limbaugh's primary targets—namely, President Clinton and government spending—as well as one item that Limbaugh became more supportive of over time—former Senator and Republican presidential nominee Bob Dole. Finally, this section looks at changes in congressional and presidential vote choices between 1994 and 1996 as a function of Limbaugh listening. If significant relationships existed between Limbaugh listening in 1995 and choices in 1996, controlling for those same choices in 1994, then the null hypothesis of no relationship between listening and choice will have absorbed a heavy blow indeed.

Table 4.4 displays the relationship between Limbaugh listening and *changes* in support for the president between 1994 and 1996. To obtain as complete and reliable a measure of support for Clinton as possible, we employed factor analysis to produce a factor score index of ten items that are repeated in both the 1994 and 1996 wave of the survey: (1) a one-hundred-point feeling thermometer, (2) whether Clinton has ever made the respondent feel "angry," (3) whether Clinton has ever made the respondent feel "hopeful," (4) whether Clinton has ever made the respondent feel "afraid," (5) whether Clinton has ever made the respondent feel "proud," (6) the degree to which the respondent believes that "moral" describes Clinton, (7) the degree to which the respondent believes that "strong leader" describes Clinton, (8) the degree to which the respondent believes that Clinton "cares" about him or her, (9) the degree to which the respondent believes that Clinton is "knowledgeable," and (10) the degree to which the respondent believes Clinton "gets things done."

In 1994, the factor displayed an eigenvalue of 4.1, explaining more than 48 percent of the variance in the individual items that made up the scale. The feeling thermometer contributed to the index most strongly, displaying a communality of .75, followed by: the degree to which one believes that Clinton "cares" (.62), the degree to which one believes Clinton exhibits strong leadership (.59), the degree to which one believes Clinton is moral (.53), the degree to which one believes that Clinton "gets things done" (.40), whether Clinton made one feel proud (.36), the degree to which one believes Clinton is knowledgeable (.34), whether Clinton inspires hope in the respondent (.32), the degree to which Clinton inspires anger in the respondent (.29), and finally, the degree to which Clinton inspires fear in the respondent (.19).

In 1996, this Clinton factor exhibited an eigenvalue of 5.1, explaining more than 50 percent of the variance in the ten variables from which it was derived. In terms of the relative contribution of each item to the index, the observed variables performed nearly equally to those in the 1994 factor.

As can be observed in table 4.4, the results indicate that listening to Limbaugh predicts changes in attitudes toward the president quite well. A 1 standard deviation in listening produces a .14 standard deviation change in warmth toward the president. In other words, a 4 standard deviation change (going from nonlistening to listening regularly) would lead to a .56 standard deviation change in attitude toward the president, an induced change of more than 15 percentile ranks. Moreover, the *t*-value indicates that the probability that these findings are a function of sampling error is less than .001.

What about policy preferences? As discussed earlier, Limbaugh emphasizes domestic economic and social welfare policy on his show. Therefore, we look at whether Limbaugh listening is associated with greater opposition over time to government spending on non-defense-related goals. We capture attitudes toward government spending in a single factor score index that includes indicators of opinions toward federal government spending in general, as well as federal spending on health care, the environment, welfare, AIDS, social security, and public schools. I combine these variables into a single latent factor in an attempt to purge the measure of the idiosyncratic features of each observed indicator (e.g., antigay sentiment driving opposition to AIDS spending), in order to create a more reliable measure of latent attitude toward spending in general. Using principle axis extraction, the factor analysis extracted a single factor (eigenvalue = 3.01). The factor explains 43 percent of the variance in the seven observed variables. Concerning each observed indicator individually, the factor explains 34 percent of the variance in attitudes toward environmental spending, 32 percent of the variance in attitudes toward welfare spending, 40 percent of the variance in AIDS spending, 24 percent of the variance in Social Security spending, 31 percent of the variance in public school spending, 23 percent of the variance in health care spending, and 53 percent of the variance in general preferences for less government services and spending.

As columns six and seven of table 4.4 demonstrate, Limbaugh listening in 1995 significantly predicts increases in opposition between 1994 and 1996 to government spending. This does not, of course, definitively demonstrate that Limbaugh persuades listeners to hold more-conservative opinions, but it supports that hypothesis, showing that some attitudes did change, and that those changes correspond to increased listening.

Support for Dole

But is the association between listening to Limbaugh and increased conservatism restricted to encouraging hostility, or can regular exposure to Limbaugh lead to greater positive evaluations of items that Limbaugh supports? Furthermore, when Limbaugh changes his tune toward a particular individual or institution, do listeners follow suit? Columns four and five of table 4.4 show the relationship between Limbaugh listening and change in support for Bob Dole. As noted earlier, research by Jones (1997) and others found that Limbaugh's message regarding Dole changed from ambivalence to full-blown support as the presidential campaign wore on in 1996. As such, examining changes in support for Dole provides a nice case study of whether Limbaugh may induce positive sentiment.

We measure support for Dole with one-hundred-point feeling thermometers in 1994 and 1996. As with the examination of attitudes toward Clinton and government spending, we include the 1994 measure as a control variable in the model predicting 1996 attitudes. However, unlike the previous panel analyses described in this section and depicted in table 4.4, we do not measure Limbaugh listening in 1995 only, because Limbaugh's strong support for Dole did not occur until roughly midway through 1996. Instead, we measure Limbaugh listening with a three-point ordinal scale, in which 0 = nonlistening in both 1995 and 1996, 1 = regular listening (at least once a week) in 1995, and 2 = regular listening in both 1995 and 1996. This measure is not as pure as the 1995-only measure, because listening in 1996 could be a function of enhanced support for Dole, rather than the reverse. However, although we approach the results obtained with this measure with greater caution, we believe that it is the best measure among imperfect alternatives.

Did listener opinion reflect the change in Limbaugh's message between 1994 and 1996? As table 4.4 reveals, although the relationship is not nearly as pronounced as when Limbaugh encourages negativity, this analysis appears to indicate that regular exposure to Limbaugh did lead to greater support for Dole over time. These results suggest that while Limbaugh appears to be quite able to induce listeners to become more hostile toward particular ideas, personalities, and groups, he perhaps enjoys a modicum of success when trying to mobilize support for particular candidates.

Vote Choice

Do these changes in opinion translate into changes at the ballot box? Because vote choice is measured dichotomously, we employed logistic regression to predict presidential and House voting decisions in 1996 by the level of Limbaugh listening in 1995, controlling for vote choices (Republican or Democrat) in 1994. Regarding the House vote in 1996, table 4.5 reveals the efficacy of Limbaugh listening as a voting determinant. A one-unit change in listening leads to an increase of 56 percent in the odds that one would have voted for a Republican member of Congress. In other words, a four-unit change (listening to Limbaugh every day) made one 5.92 times as likely to change his or her House vote from Democratic to Republican in 1996. In terms of statis-

Table 4.5 Logistic Regression of Vote Change from 1994 to 1996 on Limbaugh Listening

Independent variables	President		House	
	Logit b (exp. of *b*)[a]	**std. error**	**Logit b (exp. of *b*)[a]**	**std. error**
1995 Limbaugh listening	.51 (1.67)	.20***	.45 (1.56)	.20**
1994 House vote	2.01 (7.45)	.35***	2.42 (11.25)	.35***
Constant	−3.49	.60	−3.90	.58
R^2	.33[b]		.50[b]	
% reported	72		78	
N	197		203	

Source: American National Election Study, 1994–95 panel

a Exp.-b equals the increase in the odds of a Republican vote for a one-unit increase in the independent variable.

b Nagelkerke R2

** $p < .05$, one-tailed test

*** $p < .01$, one-tailed test

tical significance, the probability that this finding is a function of sampling error is less than 5 percent.

Regarding the presidential election of 1996, a one-unit change in listening produces a 67 percent increase in the odds that one would have voted for Dole. Hence movement from nonlistening to listening every day (four points on the scale) made one 7.78 times as likely to vote for Bob Dole, controlling for whether one voted Republican in 1994.

Conclusion

These analyses have revealed an unmistakable pattern: When Limbaugh levels criticism toward particular ideas, groups, or individuals on at least half of his broadcasts, regular listeners show a marked tendency to "buy" the Limbaugh message—displaying hostility toward those items beyond what can be accounted for by ideology, party identification, exposure to other conservative messages, affect for Limbaugh, or a host of other factors. Moreover, regular listening not only correlates with attitudes that reflect Limbaugh's message; listening also leads to opinion change toward greater conservatism and antipathy toward Limbaugh's favorite targets.

Although even panel data cannot determine causality with certainty, these results lend much credence to hypotheses positing a "Limbaugh effect." However, Limbaugh must work hard for this apparent persuasion. The less he discusses an issue, the less influence he appears to have. For issues that Limbaugh discusses on less than one-fifth of his broadcasts, we found no independent relationship between listening and opinion. Therefore, it appears that listening to Limbaugh does not produce ideological "spillover" or learning across the board.

Furthermore, Limbaugh appears to have much less success mobilizing support than in mobilizing opposition. In our cross-sectional analysis, we did not find a substantial independent relationship between regular listening and positive feelings toward anyone or anything. When we examined opinion change, we did observe a significant increase in support for Bob Dole between 1994 and 1996, corresponding to Limbaugh's change of tune toward Dole during 1996. But this increase in positivity paled in comparison to the increase in listener hostility toward other items over the same time period. Finally, listening to Limbaugh in 1995 corresponds to a significant change in the odds of voting Republican between 1994 and 1996.

It is unclear whether Limbaugh's apparent persuasion affects deep-seated attitudes, or whether it is simply a capricious response to repeated priming on the part of Limbaugh. However, the potential implications for politics may be the same regardless. Roughly one-fifth of American voters in 1994 listened to Limbaugh on a regular basis. New media such as talk radio may have reincarnated the partisan press of the nineteenth century, trading ink for airwaves. As a greater proportion of the public obtains political information via such nontraditional sources, the dynamics of media influence in the realm of public opinion may be fundamentally and immutably altered.

Chapter 5 continues to examine the relationship between Limbaugh listening and opinion, but does so within the context of the 2000 Republican primary electoral contest. The chapter also examines how variance in listener sophistication may affect the persuasion process.

5 Talk Radio, Opinion Leadership, and Presidential Nominations: Evidence from the 2000 Republican Primary Battle

Chapter 4 explored whether habitual consumption of "new media" may result in induced political choices. Specifically, I examined the extent to which the empirical association between regular listening to the Rush Limbaugh radio program and political conservatism can be attributed to persuasion effects. The primary competing explanation, of course, is that the causality is reversed—listening is a function of conservatism, not the other way around. Results provided strong support for the notion that opinion leadership over the radio airwaves is real; it seems that Limbaugh does induce his audience to be more conservative over time. But many observers might object to the generalizability of those findings, arguing that talk radio was at its zenith in 1993–96 and may have faded somewhat over time in terms of audience or influence. This chapter primarily serves to update the analysis of the previous chapter by examining opinion during the 2000 presidential nomination campaign.

Analyzing only Republicans, I observed how variance in Limbaugh listening corresponded to the likelihood of preferring John McCain to George W. Bush. The competitive primary battle between the two candidates provided an ideal setting to test for opinion leadership on the part of Limbaugh. By limiting the analysis to Republicans choosing among their own, I was able to employ another method of battling the effects of selection bias, because virtually everyone who would avoid the Limbaugh radio show for ideological reasons was effectively removed from the analysis. Perhaps of more importance, the 2000 primary battle and Limbaugh's unabashed support for Bush/opposition to

McCain allowed me to assess whether "new media" effects in the form of talk radio persuasion were limited to a unique time period (the early to mid-1990s). From an applied perspective, I was also able to observe the degree to which the specific medium of talk radio still has any marketing swagger in the marketplace of political information. Finally, this analysis seeks to gain understanding of a fascinating, ephemeral, but less-understood (and still relatively understudied) component of American presidential selection—the nomination process. In particular, I hope to offer a complement to the momentum-driven explanations of nomination outcomes (e.g., Bartels 1988) that rely on the role of partisan opinion leadership. I essentially argue that even though party labels are removed from candidates in nomination elections, and the electorate is usually limited to voters of a single political party, that partisanship still matters tremendously, and that party leadership still determines the outcome of presidential nominations, even in the absence of "smoke-filled rooms."

Vote Choice in Primary Elections

The mechanics of presidential nominations have changed dramatically over the last thirty years. Following the tumultuous Democratic National Convention in 1968, when the American public was treated to images of protestors being beaten with billy clubs by police outside the convention hall in Chicago, which was reflective of the dissatisfaction that many Americans felt toward the democratic efficacy of our national institutions, the Democratic party created the McGovern-Fraser Commission to review the ways that delegates were selected to the national convention and propose revisions. Though an unintended consequence, the reforms implemented as a result of the commission's proposals resulted in the explosion of binding primaries, where delegates are selected who are committed to the candidate winning the popular vote in a given state's primary or caucus election. This sea change in the way candidates are nominated created much confusion on the part of party officials, because the standard rules by which they had lived and grown comfortable were no longer applicable. Many lamented the erosion of party influence over the process, and feared that candidate quality would disintegrate in the hands of a selection process determined by the masses.

But while party organizations no longer enjoy direct control over the nomination process, to draw the conclusion that they no longer determine their party's standard bearer may be premature. Short of stuffing ballot boxes, how can parties exercise influence over the decisions that the rank-and-file voters make? I argue that parties exercise such influence through opinion leadership. That is, while voters in primary elections do not enjoy the simple party-identification heuristic that enables them to make "satisficed" choices in other types of elections, voters can still rely on cues (usually not so subtle) offered by prominent party leaders about who among the field of contenders represents the strongest and most viable candidate. Such partisan opinion leaders may represent governors, members of congress, a sitting president or former president, celebrities, or visible partisans who hold forth from new-media pulpits such as Internet sites, television talk shows both political *(O'Reilly Factor, Crossfire)* and pseudopolitical *(700 Club)*, and, of course, talk radio.

This view of partisan opinion leadership represents a departure from the dominant model of primary vote choice, which relies on mechanisms of "momentum" to explain primary electoral outcomes (Bartels 1988). Momentum is reflected in the ability of candidates to translate strong showings in early primary contests into greater support in later contests (for an excellent analysis of the actual mechanisms of such momentum, see Mutz 1997). One explanation argues that voters come to perceive a particular candidate or candidates more favorably based on the perception that other voters like themselves prefer a particular candidate or candidates. An alternative explanation posits that early success is translated into greater attention from the mainstream press, and this increase in visibility breeds more success. I do not argue that such momentum effects are not a central determinant of vote choice in primary elections. Rather, I argue that, for partisan voters, perceptions of the preferences of trusted party leaders may carry more weight in the minds of primary voters than do the preferences of fellow ordinary citizens or journalists. Moreover, I contend that momentum itself may stem in part from partisan opinion leadership, particularly in contests where the party is unified behind a single candidate (e.g., 1996, 2000), and in states where scheduling precludes the ability on the part of most voters to obtain adequate "face time" with the candidates themselves (everywhere except New Hampshire and Iowa).

Thus this study follows on the heels of the massive analysis undertaken by Zaller and his colleagues that argues that party leadership much more efficiently explains primary outcomes in the modern era than does simple momentum, for every year other than 1976 (Zaller 2001a). This study attempts to provide a model of how that party influence becomes manifest in the form of opinion leadership.

On the contemporary American political landscape, few opinion leaders within the Republican Party have a mouthpiece the size of Rush Limbaugh's. Although not quite the fad or cultlike figure he was in the early 1990s, his show is still broadcast five days a week, three hours a day on more than six hundred stations nationwide and, according to the survey conducted for this analysis (2000), still commands the attention of roughly one-fourth of Allegheny County Republican voters at least twice a week (even in a northern Rust Belt city known as a union stronghold). The following section will briefly review the circumstances of the 2000 primary battle between John McCain and George W. Bush, pointing to Bush's position as "favorite son," and Limbaugh's role as party mouthpiece, out in front of the collective Republican establishment's march to quell the McCain insurgency.

The Struggle for the 2000 Republican Presidential Nomination

The battle for the Republican nomination for president between George W. Bush, the front-runner, and John McCain, the challenger, proved to be one of the most memorable nomination campaigns in recent political history. Not since Reagan challenged Ford for the 1976 nomination has an insurgent Republican garnered as much media attention or caused as much hand-wringing on the part of the party's front-runner. Although Bush effectively secured the nomination on 7 March, barely five weeks into the official primary season, this seemingly quick victory masks what was a highly contentious contest that had been very much in doubt after McCain stunned Bush on 1 February with a nineteen-point victory in New Hampshire—an embarrassment to the well-connected candidate who had raised a record $60 million in campaign contributions during the precampaign period and had lined up endorsements of nearly every Republican governor, congressperson, and major party player. The momentum McCain gained from his win in New Hampshire obliterated commanding leads that Bush had enjoyed in public opinion polls in places such as South Carolina, Michigan, and New York.

But as the Bush campaign struggled, various conservative groups and party leaders who had committed publicly to Bush, and therefore had some stake in his political fortunes, began working in earnest to sabotage the McCain insurgency (Chaitt 2000). No one was more active in this endeavor than Limbaugh, who accused McCain of "dividing people" and engaging in "Clintonesque exploitation par excellence" (Hibbs 2000). Limbaugh went so far as to suggest (Hibbs 2000) themes for campaign commercials for the South Carolina contest that would portray McCain supporters as wolves in sheep's clothing—closet Gore supporters hell bent on derailing the Bush candidacy (Edsall and Neal 2000).

After a hard-fought and decidedly nasty campaign in South Carolina, Bush rebounded with a decisive win, only to be upstaged again three days later in Michigan and McCain's home state of Arizona. Then McCain committed perhaps the most egregious tactical error of his campaign, delivering a speech near Pat Robertson's hometown that called the religious leader "evil" and an "agent of intolerance." This statement was interpreted as an assault on the Christian Right, who make up the most reliable voting bloc in the Republican Party (Barker and Carman 2000). Bush took the Virginia primary convincingly, receiving well over 90 percent of the evangelical Christian vote, and the McCain Express began slowing. It ultimately ground to a halt after Super Tuesday.

The primary charge leveled at McCain by the Republican establishment was that he was no longer a "true Republican" (Chaitt 2000). From campaign finance reform to tobacco regulation to tax reform, McCain found himself at odds with Republican congressional leaders and conservative media professionals concentrated within new media outlets (such as Pat Robertson, George Will, Bill Bennett, Sean Hannity, and Bill O'Reilly; see Chaitt 2000). This suspicion of McCain on the part of conservative leaders may be summed up in a statement offered by Limbaugh: "When I hear McCain using liberal rhetoric to bust up the conservative coalition, I think, what the hell is this. . . . This guy's a Republican . . . I'm just an honest-to-God thoroughbred conservative, and I don't see McCain as that" (Hibbs 2000).

What effect did Republican Party stumping for the candidacy of George W. Bush have on voters? The following section analyzes the relationship between listening to Republican bullhorn Rush Limbaugh and various political attitudes during the 2000 primary campaign season, including Republican voter preference between Bush and McCain.

Research Design and Methodology

To determine the relationship between Limbaugh listening and primary vote choice, I analyzed data from a CATI (computer-assisted telephone interviewing) survey of 287 registered Republican voters in Allegheny County, Pennsylvania (which comprises the city of Pittsburgh and its surrounding area). The data was collected by the University Center for Social and Urban Research at the University of Pittsburgh during January, February, and March 2000 (ending prior to the Pennsylvania primary contest).

The Sample

As mentioned earlier, these Allegheny County Republicans appeared, on the whole, to be well acquainted with Limbaugh. While 66 percent of the sample indicated that they never listened to Limbaugh's show, 34 percent said they listened at least once a week, 25 percent said they listened at least twice a week, and 10 percent said they listened nearly every day. Furthermore, the sample was overwhelmingly white (93%), well educated (47% with bachelor's degrees), mature (mean age was 56), and married (75%), but representative of the population in terms of gender (51% female). Twenty-three percent of the sample identified themselves as "born-again Christians."

On the whole, the sample was relatively politically astute. Using a modified form of Delli Carpini and Keeter's Knowledge Index (1996), I found that my Republican sample was substantially more likely than Delli Carpini and Keeter's sample to know (1) which party controls the House of Representatives (72% to 55%), (2) which party is more conservative (88% to 57%), (3) whose responsibility it is to determine the constitutionality of a law (82% to 68%), and (4) how much of a majority is needed in Congress to override a presidential veto (57% to 37%). Moreover, 58 percent of the sample knew which office was held by Madeleine Albright, and 72 percent were able to correctly identify the mayor of New York City (Rudy Giuliani).

Dependent Variables

There are two categories of choices that I attempted to explain from this survey analysis: liberalism-conservatism and candidate preference. As with chapter 4, I expected the magnitude of the relationship between Limbaugh listening and liberalism-conservatism to depend on the policy area. Given that Limbaugh emphasizes discussion of domestic government spending and other economic issues at the expense of cultural issues (such as gay rights and abortion) and foreign policy issues (see chapter 2), I expected to find robust independent relationships between Limbaugh listening and variables pertaining to economic liberalism-conservatism while observing weak relationships between Limbaugh listening and variables pertaining to foreign policy or cultural liberalism-conservatism.

Economic liberalism-conservatism was measured in two distinct ways, one policy based and the other value based. First, I measured policy-based economic liberalism-conservatism by asking respondents to indicate a preference for the degree to which the federal government should spend money and provide services. Readers familiar with the American National Election Studies (ANES) surveys will recognize this measure, which has been used by the ANES as a measure of economic ideology for several consecutive electoral cycles and has consistently held up to scrutiny regarding the validity and reliability of the measure. Responses were coded on a seven-point scale, ranging from −3 to +3, where negative numbers reflected conservative preferences for less services/spending, with preferences for much less spending and many fewer services coded as "−3."

The value-based measure attempted to capture the same underlying ideology without reference to particular policies. It is essentially a measure of individualism-communitarianism; respondents were asked which of the following principles was more important for society: "self-reliance and personal responsibility, or cooperation and helping others." Responses were dichotomous—respondents were forced to make a choice between the competing values. Conservative responses (preferences for self-reliance and personal responsibility over cooperation and helping others) were coded as "0"; liberal responses were coded as "1."

Cultural liberalism-conservatism was captured with a single indicator that measured respondents' attitudes toward "new lifestyles." Respondents were asked whether new lifestyles were good, bad, or neither good nor bad for society. This measure was deliberately chosen for its vagueness. "New lifestyles" may certainly refer to several different things, ranging from single-parent families, the perceived breakdown of respect for authority, illicit drug use, homosexual couples, nontraditional gender roles, or even rampant narcissism and self-indulgence. Surely, different respondents interpreted the question in different ways.

But nearly every interpretation of "new lifestyles" refers to the perceived breakdown of traditional "family values," which cultural conservatives hold to be sacred and which form the center of the cultural conservative agenda. As mentioned earlier and elaborated on in detail in chapters 2 and 4 of this book, Limbaugh's message is not centered on discussions of family values or new lifestyles, so I do not expect to find a significant independent relationship between Limbaugh listening and cultural conservatism among registered Republicans.

Foreign policy attitudes are captured in a single indicator of isolationism-internationalism. Respondents were asked which was more important for America: "avoiding international conflict, or protecting our interests worldwide?" One half of the respondents were randomly queried with options in the reverse order, in order to eliminate bias toward selecting the first option. "Avoiding international conflict" was coded as "0," while "protecting our interests worldwide" was coded as "1." Again, because Limbaugh spends little time discussing foreign policy, I do not expect to find an independent relationship between listening and isolationism-internationalism. Both this measure and the cultural ideology measure merely serve as points of comparison, to better gauge the meaningfulness of the hypothesized relationship between listening and economic ideology.

Finally, voters were asked which of the Republican candidates they preferred, among Bush, McCain, Forbes, Hatch, Bauer, and Keyes. As expected, the Republican sample was partial to Bush, preferring him to McCain by nearly 19 percentage points (49% to 30%). However, the strong preference for Bush is somewhat inflated due to the overwhelming support that he enjoyed during January, when few Pennsylvania Republicans had yet acquainted themselves with the McCain candidacy in a meaningful way. Twenty-one percent of the sample preferred a different candidate or were unable/unwilling to indicate a preference.

Formally stated, the hypotheses tested in the first part of this chapter are as follows:

H1: The more frequently a Republican respondent reported listening to Rush Limbaugh in early 2000, the more likely that respondent was to prefer fewer government services and less government spending.

H2: The more frequently a Republican respondent reported listening to Rush Limbaugh in early 2000, the more likely that respondent was to see the values of self-reliance and personal responsibility as more important than cooperation and helping others.

H3: The more frequently a Republican respondent reported listening to Rush Limbaugh in early 2000, the more likely that respondent was to support George W. Bush in the 2000 primaries.

Independent and Control Variables

To test these hypotheses, I performed a series of ordinary least squares and logistic-regression analyses. To isolate the relationship between listening and political choices and to ward off spurious relationships, I included the following control variables in the regression equations: the date the respondent was interviewed (to measure momentum); traditional media usage (an index of the number of days per week that the respondent watches national news, local news, and reads a newspaper); Internet usage for political information; political knowledge (measured with a factor score index of six items: [1] Which party controls the U.S. House? [2] Which party is more conservative? [3] Which political office is held by Madeleine Albright? [4] Whose responsibility is it to determine if a law is constitutional or not? [5] How much of a majority in Congress is needed to override a presidential veto? [6] Who is the mayor of New York City?—the Cronbach's alpha for the index was .71; one factor was extracted using principal axis extraction, indicating that this index is a valid and reliable measure of political knowledge); whether a respondent was a born-again Christian; gender; level of educational attainment; income; age; marital status; which candidate the respondent believed was more likely to defeat the Democratic nominee; and which candidate the respondent believed the media wanted to win the nomination.

Findings

Table 5.1 reports the findings from these regression models. The second column displays the results of the model predicting attitudes toward federal government services and spending. As expected, Limbaugh listening is strongly associated with political individualism as expressed in the form of opposition to federal government activity in the economy. The strength of listening is impressive, given that everyone in the sample is a registered Republican and therefore more inclined than the average American to oppose federal government intervention. Perhaps not surprisingly, given the Republican-only sample, very little else

Table 5.1 Prediction Models of Primary Voter Choice: Ideology and Candidate Preference, 2000

Dependent Variables:	Govt. Role[1]	Individualism[2]	New Lifestyles[3]	Foreign Policy[4]	Bush v. McCain[5]
Estimation Method:	OLS	Logit	OLS	Logit	Logit
Coefficient:	*b*	Δ in odds ratio[6]	*b*	Δ in odds ratio[6]	Δ in odds ratio
Independent variables:					
Limbaugh	-.06*	.82*	-.00	.98	.55*
Media consumption	.01	1.03	-.00	1.03	1.07
Internet use	.00	1.05	-.00	1.05	1.07
Year born	-.00	1.01	-.00	1.02*	1.02
Income	-.03	.76**	-.00	.92	1.38
Education	-.06	1.18	.02	1.18	.54*
Female	-.27*	1.16	.01	.54*	.77
Born-again	-.01	1.06	-.00	1.64	.32
Political knowledge	-.10	.33***	-.00	.86	1.94
Interview date	-.00	.99	-.00	1.00	1.00
McCain personal					1.06
Strategic choice					.02***
Perceived media bias					3.81*
Govt. role					4.73***
New lifestyles					3.96*
PRE statistic	.10	.33	.06	.09	.76
N	287	287	287	287	287

Source: 2000 Allegheny County survey of Republican primary voters

[1] This dependent variable is coded from 1 to 7 such that opposition to federal spending and services is coded with lower numbers and support for such governmental activity is coded with larger numbers.

[2] This dependent variable is coded dichotomously, such that a preference for cooperation and helping others equals "1," while a preference for self-reliance and personal responsibility equals "0."

[3] This variable is coded from 1 to 3 such that "1" equals a belief that new lifestyles are bad for society, and "3" equals a belief that new lifestyles are good for society.

[4] This variable is coded dichotomously so that "0" equals a preference for isolationism and "1" equals a preference for international interventionism.

[5] This dependent variable is coded dichotomously, such that a preference for McCain equals "1," while a preference for Bush equals "0."

[6] The change in the odds ratio is the anti-log of the logistic regression coefficient. It represents the change in the odds of an event occurring for a one-unit increase in the independent variable. Coefficients less than one represent negative relationships. Coefficients close to one represent very small substantive relationships. In this table, a one-unit increase in Limbaugh listening (an increase in listening by one day per week) corresponds to an 18% increase in the likelihood that a respondent will prefer self-reliance to helping others, a 2% increase in the likelihood of preferring isolationism to interventionism, and a 45% increase in the likelihood of voting for Bush.

* $p < .05$
** $p < .01$
*** $p < .001$

significantly predicts respondent attitudes toward spending/services. The "usual suspects," which typically predict such sentiment in samples that are not restricted by partisanship—such as income and education—are not statistically significant in this model. Moreover, measures of other types of media consumption (traditional media usage, Internet usage) fail to predict spending/services attitudes in a meaningful way. One interesting finding worth noting is the strong association between gender and spending attitudes. Surveys often find gender effects relating to public opinion, but males are typically more conservative economically than females—when considering the population as a whole. Here, I find that when restricting the analysis to registered Republicans only, women tend to be more conservative economically than men, at least in this community study. Further research using a national survey representative of the entire American electorate should be conducted to examine whether Republican women tend to be more economically conservative than Republican men.

Column 3 of table 5.1 provides further support of the empirical relationship between Limbaugh listening and political individualism by examining political individualism from a pure value-based perspective rather than a policy-based perspective. The logistic-regression equation reveals that a full range change in Limbaugh listening—that is, Republicans who listen five days a week, compared to Republican nonlisteners, are more than four times as likely to indicate that self-reliance and personal responsibility are more important principles on which to base society than are cooperation and helping others. Not surprisingly, personal income is also a significant predictor. It is interesting to note that political knowledge is also related to individualism in this model, indicating that those more politically astute are more likely to adhere to "conservative" principles. This follows intuition, given the makeup of the sample; sophisticated Republicans are more likely to draw connections between their partisanship and the philosophical underpinnings of the ideology for which the party stands.

Columns 4 and 5 serve as points of contrast to columns 2 and 3. These columns predict attitudes toward issues that are not often addressed on Limbaugh's broadcasts, but nevertheless represent salient issues of contention within contemporary American politics, which can be explained well by ideology and party identification within the general population (at least in the case of cultural issues). Column 4 predicts attitudes regarding whether "new lifestyles are good for society," while column 5 predicts attitudes pertaining to how active the United

States should be on the world stage. Conservatives are typically more likely to believe that "new lifestyles" are bad for society. Foreign policy attitudes are more complicated, but since the Vietnam era, liberalism has been associated with opposition to military intervention abroad as well as opposition to free trade. As can be observed in table 5.1, neither cultural attitudes nor foreign policy attitudes can be explained in any meaningful way by the degree to which a respondent listens to Rush Limbaugh. This lack of a statistically significant relationship does not mean that Limbaugh listeners are not social conservatives or foreign policy conservatives. Rather, it indicates that Limbaugh listeners are, on average, no *more* conservative on these issues than are other rank-and-file Republicans. This lack of a relationship regarding issues largely ignored by Limbaugh provides further support for our hypothesis that the relationships that are present regarding issues that Limbaugh does address are not entirely a function of self-selection on the part of listeners, but rather constitute an example of political persuasion and opinion leadership by a partisan mouthpiece.

The sixth column of table 5.1 provides perhaps the best evidence yet of opinion leadership over the talk radio airwaves. Predicting candidate preference among Republican voters between two competing Republican candidates with similar policy stands and records provides something of a natural experiment, where the prospect of listeners selecting the Limbaugh show on the basis of an expectation that Limbaugh likes Bush seems illogical. Therefore, the fear of selection bias contaminating our prediction model is minimized. The analysis of primary vote choice is also relevant because primaries are now a central component of the American democratic process that is nevertheless still something of a mystery to political scientists. Not only may the study of primary vote choice models signify the practical, "real-world" relevance of talk radio in a way that the analysis of public opinion or general elections cannot, but, as mentioned at the outset of this chapter, talk radio influence in this context may provide support for an alternate theory of primary electoral outcomes—one that depends more on party leadership than on strategic voting, momentum, or "bandwagon" effects.

This model adds a couple extra control variables, to further isolate any talk radio effect from alternate hypotheses. First, I added a variable measuring the degree to which a respondent likes McCain in terms of his personal characteristics. This variable is an index comprising several individual assessments of how well particular adjectives described McCain.

The adjectives were "honest," "caring," "principled," "smart," visionary," and "boring." Also, I added the policy variables that were examined as dependent variables from the previous analyses (spending preferences and attitudes toward new lifestyles) as control variables, to further control for any ideological influence that restricting the sample to Republicans only may have left present. Thus I have attempted to create a model biased against finding a Limbaugh effect, reasoning that type II statistical errors (those occurring when a researcher falsely concludes that there is no causal relationship between variables) are preferable to type I errors (when a researcher falsely concludes that a causal relationship is present). Hence conservatism in model specification was the order of the day for this analysis.

As revealed by the pseudo *R*-square statistic, this model of primary vote choice performs well, reducing the errors in prediction by nearly 74 percent and correctly predicting 88 percent of all respondent vote choices in the sample. As expected, Limbaugh listening is strongly related to vote choice. As the change in the odds ratio shows, a one-day-per-week increase in Limbaugh listening is associated with a 45 percent increase in the likelihood of voting for Bush. Thus voters who listen five days a week are 3.35 times as likely to prefer Bush than Republican voters who never listen to Limbaugh.

This apparent Limbaugh effect stands in contrast with the apparent lack of an effect of traditional media consumption. Given the presumed media bias toward McCain, which was trumpeted loudly by many during the campaign (e.g., Hibbs 2000), one might have expected a traditional media effect. If the media were biased toward McCain, such bias did not appear to translate into anything more than a "minimal effect" at best—at least among Republicans. Perhaps this can be explained by the general perception by Republicans that traditional media are biased against them (see Limbaugh 1993). I did find evidence that such a perception was strongly related to vote choice, such that people who believed that the media were biased toward McCain were nearly four times as likely to support the Bush candidacy.

In terms of the influence of other predictor variables, candidate viability in the fall election was a strong predictor of vote choice, although the causality of such a relationship is undetermined and may reflect a certain degree of projection on the part of the respondents. Finally, the ideology measures performed well, justifying their inclusion as control variables and making the robustness of the Limbaugh effect that much more impressive.

Sophistication

How is the apparent Limbaugh effect influenced by voter sophistication? Traditional media and campaign effects literature has lamented the fact that media effects may be hard to find because those who are most likely to be exposed are the least susceptible to persuasion, due to higher-than-average levels of political sophistication. However, the theoretical and empirical models of chapter 3 suggest that Limbaugh's brand of propaganda, which relies heavily on value heresthetic to encourage periphery-route persuasion, may actually make persuasion more likely among sophisticated audiences. This section reports findings from models including an interaction term, multiplying the political knowledge factor score by Limbaugh listening, to see if higher levels of sophistication on the part of listeners either magnifies or attenuates the relationship between listening and choice. In other words, are sophisticated listeners either more or less susceptible to Limbaugh's persuasive attempts? Given that sophisticated voters are more likely to draw connections between ideology, partisanship, and vote choice, I hypothesize that Limbaugh's appeal, which was strongly partisan—centering on the accusation that McCain was not a "real" Republican—would resonate more strongly among listeners who had a strong sense of partisanship. Formally:

H4: The relationship between Limbaugh listening and political choice, in terms of public opinion or vote preference, becomes stronger as political knowledge increases.

Table 5.2 displays the results of the interaction models predicting attitudes toward services/spending, self-reliance versus humanitarianism, and vote choice. Each of the regression equations mirrors those in table 5.1, with the addition of an interaction term multiplying Limbaugh listening times knowledge. In two of the equations—those predicting spending/services attitudes and vote choice—sophistication clearly seems to matter. Looking at the coefficients associated with the Limbaugh variable in isolation and with the interaction term, one observes that while Limbaugh listening bears a relationship to economic conservatism and Bush support, these relationships become much more pronounced when listening is combined with higher levels of political knowledge. In this case, it seems that greater sophistication does not inoculate one from being influenced by partisan propaganda. Conversely, sophistication does not seem to alter the relationship between Limbaugh listening and

Table 5.2 Prediction Models of Primary Voter Choice: The Interaction of Limbaugh Listening and Political Knowledge

Dependent Variables:	Govt. Role[1]	Individualism[2]	Bush v. McCain[3]
Estimation Method:	OLS	Logit	Logit
Coefficient:	*b*	Δ in odds ratio[4]	Δ in odds ratio
Independent variables:			
Limbaugh	−.05	.83	.60
Limbaugh × knowledge	−.14**	.93	.27**
Media consumption	.01	1.03	1.18
Internet use	.00	1.05	.92
Year born	.00	1.01	1.05
Income	−.00	.78***	1.48
Education	−.07	1.16	.44**
Female	−.27*	1.17	1.07
Born-again	−.02	1.05	.34
Political knowledge	−.01	.34***	3.70*
Interview date			.99
McCain personal			1.09
Strategic choice			.01***
Perceived media bias			3.56
Govt. role			6.60***
New lifestyles			5.58**
PRE statistic	.12	.33	.79
N	287	287	287

Source: 2000 Allegheny County survey of Republican primary voters

[1] This dependent variable is coded from 1 to 7 such that opposition to federal spending and services is coded with lower numbers and support for such governmental activity is coded with larger numbers.

[2] This dependent variable is coded dichotomously, such that a preference for cooperation and helping others equals "1," while a preference for self-reliance and personal responsibility equals "0."

[3] This dependent variable is coded dichotomously such that a preference for McCain equals "1," while a preference for Bush equals "0."

[4] The change in the odds ratio is the anti-log of the logistic regression coefficient. It represents the change in the odds of an event occurring for a one-unit increase in the independent variable. Coefficients less than 1 represent negative relationships. Coefficients close to 1 represent very small substantive relationships.

* $p < .05$
** $p < .01$
*** $p < .001$

value preference. Perhaps this is because more political knowledge is required to form a policy preference than a basic value orientation.

Discussion

This chapter has examined the role of partisan opinion leadership on citizen vote choice in presidential nomination elections by observing the relationship between Limbaugh listening and voter choice in the 2000 nomination contest between John McCain and George W. Bush. This particular contest served as an ideal case study because Republican Party leaders were nearly universally supportive of the Bush candidacy, and were thus engaged in quelling the McCain insurgency. Also, by analyzing choice in a nonpartisan election in 2000, this chapter serves to update and extend the assessment of talk radio as a choice determinant that had been the focus of the previous two chapters. I found that the more the Allegheny County Republicans listened to Rush Limbaugh, the more likely they were to express economically conservative attitudes, both in terms of policy preference and value orientation. Greater Limbaugh listening was also strongly associated with a preference for Bush over McCain, controlling for strategic thinking, traditional media consumption, attitudes toward McCain's personal characteristics, momentum, and traditional demographics. These findings were expected because of Limbaugh's commitment to economic individualism and his vitriolic opposition to the McCain candidacy. Conversely, I found no independent relationship between Limbaugh listening and ideology regarding issue realms that are not emphasized on Limbaugh's show, including cultural issues and foreign policy. These findings serve to corroborate the findings from chapter 4, and indicate that the "Limbaugh effect" was not restricted to the mid-1990s. More important, these findings provide support for the theory that partisan leadership may be more central to determining presidential nominations than is often assumed in the post-convention-primary era. It appears that party leaders may have considerable influence over the choices that the rank and file make, particularly if they convince voters that a particular candidate is not true to the preferred partisan ideology or represents a threat to the party. Such renewed party leadership may be possible because of the growth in new media, where partisan political personalities are able to reach large audiences directly, in ways that were much more difficult twenty years ago. A model of determining nomination outcomes that heavily depends upon the role of parties challenges conventional wisdom—which points to candidate momentum (as gained through early victory, increased media exposure, and bandwagon effects) as the primary nominating determinant (e.g., Bartels 1988).

Finally, this chapter sought to understand how voter sophistication may interact with exposure to messages to influence the persuasion process. I found that Limbaugh listening bore an even more dramatic relationship to economic policy preference and primary vote choice for politically knowledgeable voters than for the less informed. This finding provides some additional empirical support for the theoretical model of heresthetic persuasion effects that I outlined in chapter 3, where voter sophistication actually aids the persuasion process—instead of inhibiting it, as has generally been shown in political persuasion studies that rely on rhetorical efforts at persuasion (e.g., Zaller 1992).

The next chapter also considers the degree to which Limbaugh's message is persuasive. But instead of considering persuasion as it pertains to policy preferences, candidate evaluations, or vote choices, chapter 6 conceptualizes persuasion as mobilization, or induced activity. Specifically, I observe the relationship between listening to Limbaugh and levels of political participation.

6 The Talk Radio Community: Nontraditional Social Networks and Political Participation

Governmental outputs often reflect expressions of popular will (Page and Shapiro 1983; Wlezien 1995). But "popular will" depends largely upon the profile of the participatory public. Thus if contextual forces facilitate participation by individuals of one ideological bent while stifling participation by others, then the public profile becomes lopsided. As a consequence, governmental outputs will disproportionately reflect the more-participatory ideology. This chapter examines persuasion as it relates to influencing individual participation. Much research has demonstrated that an individual's propensity to participate in politics is largely determined by the degree to which that person believes he or she can make a difference (e.g., Abramson and Aldrich 1982). Such belief is often referred to as political efficacy, and it has two components: internal and external. Internal efficacy is related to personal confidence and self-esteem, while external efficacy refers to belief and trust in the political system. Little research has considered whether such efficacy is subject to exogenous influence—whether and how individual efficacy can be influenced (enhanced or dulled) by outside forces. In this chapter, I test the hypothesis that by priming internal and external political efficacy in audience members, a propagandist may then persuade people of like mind to participate more, even without explicitly "calling people to action." I posit that efficacy may be primed via two distinct mechanisms.

Portions of this chapter are reprinted with permission from Barker (1998b).

First, in the direct route, a propagandist may use heresthetic to prime individuals to feel more internally and externally efficacious. Second, in the indirect route, individuals may come to feel more internally efficacious as a by-product of perceiving that their views reflect those of the majority. In such a scenario, the message *context* may be manipulated such that audience members are induced to feel as though they are part of a "club," even without any verbal attempt by the propagandist to make audience members more confident in themselves. Rather, when audience members perceive that they are surrounded by like minds, they may gain confidence. It matters not whether an audience members' views are truly in the majority. All that matters is the audience member's perception of reality.

The constructionist school of political communication posits that people construct social reality—drawing inferences about reality, rightly or wrongly, from the messages to which they are exposed. Hence the indirect model of efficacy priming, where efficacy is primed in audience members by creating a message environment where audience members infer that their political views are legitimate and popular, fits nicely within the constructionist model of media influence (e.g., Neuman, Just, and Crigler 1992).

Therefore by empirically testing both the heresthetic and constructionist routes to efficacy priming, this chapter not only seeks to extend the heresthetic model of political persuasion (see chapter 3) to the realm of political mobilization, it also examines the extent to which membership within a nontraditional social community produces change in individual political participation.

Specifically, I posit that Limbaugh directly encourages political participation by using heresthetic to give his audience a "pep talk," while the talk radio medium affects participation indirectly, with the call-in format and predominant caller profile, prompting a particular constructed reality depending upon the ideological bent of audience members. Regarding the latter, I posit that the Rush Limbaugh radio audience effectively operates like a social network. This nontraditional social network may produce informal pressure to conform to the norms of the group. Those who conform may be rewarded with psychological benefits such as feelings of community and acceptance, which may translate into greater internal efficacy.

However, this influence differs with regard to one's ideology. In this instance, the "included" group within the community is made up of self-identified conservatives and (to a lesser extent) moderates. For these members of the listening community, listening should encourage participation, due to informal pressure provided from social interaction with other participatory members of the community (Leighley 1990; Huckfeldt and Sprague 1993). In other words, I posit that the radio program provides an environment in which conservatives are joined by "kindred spirits" in the personage of Limbaugh and callers (in the case of moderates, they are surrounded by people of at least somewhat similar mind). As a result, conservative and moderate listeners become emboldened through their heightened sense of being part of the majority.

On the other hand, liberals who throw themselves into the den of the opposition may become intimidated and unsure of their own belief structures. This confusion may not lead liberals to change their mind with regard to their vote choice, but it may produce enough uncertainty to cause such liberals to conclude that both parties "don't care," or that politics is just "too complicated." Such cynicism and confusion may lead to withdrawal from participatory activities (Brody 1978; Finifter 1974; Noelle-Neumann 1984; Capella and Jamieson 1997).

But before exploring the dynamics of this "talk radio community," the following section uses experimental methods to investigate whether efficacy can be primed through the use of heresthetic, and whether such priming can result in greater political activity.

The Efficacy-Priming Experiment

This section tests the hypothesis that a message sender may successfully move message receivers to participate more in politics by directly encouraging listeners to feel more efficacious. Just as a propagandist may seek to guide opinion by priming higher-order values, such as individualism, he or she may seek to mobilize activity by telling message receivers that they can make a difference. Much scholarship has found that such political efficacy bears strong relation to the degree to which individuals participate in politics (e.g., Verba and Nie 1972; Abramson and Aldrich 1982). However, very little attention has been given to how mobilization attempts may take advantage of this relationship. Just as priming a value such as individualism may successfully induce opinion change—at least in the short run (see chapter 3)—priming the value of efficacy may be an effective means of mobilization. Unlike specific calls to action, to which audience members might respond by digging their heels in when they realize that someone is trying to get them off the couch, heresthetic does not overtly ask audience members to *do* anything. It merely prompts them to think in different terms, which may cause those audience members to believe that they are choosing to become more active by their own volition.

I tested this hypothesis by randomly exposing experimental participants to stimuli that either did or did not contain messages designed to prime personal efficacy, and then recorded the degree to which respondents shared their views and tried to persuade others during simulated legislative committee deliberations. This design is an extension of the one described earlier in chapter 3. Recall that ninety-one University of Houston undergraduates were randomly exposed to different messages by Rush Limbaugh. Eighty of those experimental subjects also participated in this efficacy-priming and mobilization experiment.[1] In addition to containing frames emphasizing the importance of freedom and self-reliance, the heresthetic stimulus described in chapter 3 also contained messages urging listeners to be confident in their beliefs because "we are making a difference!" By contrast, neither of the other two stimuli that were used to distinguish messages in the earlier experiment contained public urges of any kind. Therefore, for this mobilization experiment, the control group grew to encompass all subjects who did not receive the value heresthetic stimulus. So thirty-one of eighty subjects were randomly exposed to heresthetic designed to prime efficacy among audience members.

Following message exposure, subjects completed posttest questionnaires in which respondents reported their degree of personal political efficacy. Efficacy was measured by responses to the question: "To what extent do you agree or disagree with the following statement: 'People like me have no say in government.' " Responses were coded on a five-point Likert scale, with "1" indicating strong agreement and "5" indicating strong disagreement. After completing these questionnaires, subjects were again randomly assigned to different groups. These groups, or "legislative committees" of 6–11 members, listened to a proctor read a hypothetical piece of legislation calling for $21 billion in new federal social spending over five years.

After hearing the bill, subjects were asked to take a secret vote on passage of the bill by simply voting "yes" or "no" on sheets of notebook paper that had been provided. A proctor then collected the slips of paper and instructed the committee to begin deliberation on the bill. To encourage participation, subjects were told that if a committee member voted with the majority in the final vote (to be taken approximately fifteen minutes later), and if he or she was fortunate enough to win the random lottery after the session was over, then an extra $15 would be awarded, bringing the overall lottery award to $65. On the other hand, if a committee member were to win the lottery, but had not voted with the majority in the final vote, then that person would receive a prize of $50. I assumed that the promise of an extra $15 would encourage committee members to argue their case with more conviction than they would have in the absence of any incentive. Committee members were told that they were free to deliberate however they wished, and that they could call a final vote at any time. Thus in this experiment, the concept of participation was measured in terms of active deliberation in a simulated legislative committee. As such, the degree of deliberation on the part of individual committee members served as the dependent variable in the analysis. As a general measure of participation, degree of deliberation in simulated legislative committee surely suffers in terms of content validity. Indeed, for most citizens, participation in the political arena entails a number of activities including voting, contacting representatives, and so on. However, perhaps the most public form of participation in politics involves proselytizing, or the attempt to influence the vote choices of others. Although we readily regard our deliberation indicator as an imperfect measure of participation generally, we believe that if individuals can be induced to participate in this most public of participatory activities, then other forms of participation, such as voting, may be induced as well. Furthermore, in a laboratory setting, propensity to take part in the deliberation over a bill provided a better opportunity to assess participation than alternatives such as the decision to vote, because the choice of whether to take part in the debate did involve some opportunity cost, namely, the threat of humiliation among peers. In the lab, we could not conceive of a way to create an opportunity cost to voting on the bill, and alternative manifestations of voting were simply not viable.

Participation in the deliberation session was extensive. In each case, the proctor had to stop the deliberation and call a final vote (due to time constraints) before the committee indicated a preference to do so. Nevertheless, in each session some committee members chose to participate very little. The question is whether those who chose to participate a great deal had been more likely to hear the message designed to promote efficacy.

The degree to which each subject participated in the deliberation was determined by three coders who independently viewed the videotaped deliberation. The coders gave each of the subjects a "participation score," ranging from "0—no participation," to "3—heavy and consistent participation." Inter-coder reliability was respectable: In 78 percent of cases, all three coders gave subjects identical scores. At least two out of the three coders gave identical scores 96 percent of the time, and in only two cases did coders give subjects scores that differed by more than one value. In the case of disagreement among coders, subjects received the score that was provided by the majority of coders. In the rare instance where all three coders provided different scores, a fourth coder was employed to "break the tie." The mean rate of participation in the session was 1.86, with a standard deviation of 1.05.

The votes of the committee members, and their reasons for them, were not of interest in this experiment. The direction of attitudes had been captured in the preceding posttest questionnaires and as such, any analysis of vote choice would have been biased by the posttest responses. The committee deliberation was added to this design merely as a means of generating political discussion in order to gauge the degree to which those primed to think in more efficacious terms responded by showing more aggression during the ensuing legislative debate.

Experimental Results

If participation may be encouraged by priming the value of personal political efficacy, we should first expect to find a concrete relationship between exposure to the heresthetic stimulus and the degree to which one believes one has "a say" in government. The first column in table 6.1 displays this relationship, controlling for race, gender, exposure to the pretest questionnaire, exposure to the Limbaugh rhetoric stimulus, ideology, and attitude toward the degree to which anything ever "changes" in politics. As the unstandardized coefficient indicates, exposure to the value-priming stimulus predicts a .58 unit increase in efficacy (on a five-point scale), a relationship that is statistically significant in a one-tailed test. This finding suggests that Limbaugh succeeded in inspiring subjects, convincing them that they "could make a difference." But to what extent do such beliefs translate into a penchant for participation?

The third column of table 6.1 shows the relationship between political efficacy and activity in the ensuing legislative committee deliberation. A 1-unit increase in efficacy accounts for a .40-unit increase in propensity to participate, all else being equal, a relationship that is statistically significant at the .01 level. Therefore a 4-unit increase in efficacy (the entire scale) translates into a 1.6-unit increase in participa-

Table 6.1 Multivariate OLS Regression of Value Priming, Efficacy, and Participation

	Dependent Variables		
Independent Variables	Efficacy	Participation (1)	Participation (2)
	b (std. error)	*b* (std. error)	*b* (std. error)
Value priming	.58 (.27)*	NA	.55 (.31)*
Rhetoric	.46 (.29)	NA	.17 (.32)
Efficacy	NA	.40 (.12)**	NA
Female	−.36 (.23)	.27 (.26)	.20 (.27)
Hispanic	.51 (.35)	−.55 (.37)	−.39 (.39)
Black	.24 (.27)	−.62 (.31)*	−.55 (.32)*
Ideology	.02 (.08)	−.06 (.08)	−.04 (.09)
Change	.44 (.12)***	.06 (.14)	.21 (.14)
Pretest	.26 (.30)	−.08 (29)	−.12 (.30)
Pilot	.80 (.44)*	NA	NA
Constant	1.34 (.60)*	.91 (.65)	1.26 (.68)
N	91	80	80
Adj. R^2	.16***	.07*	.01

* $p < .05$, one-tailed test
** $p < .01$, one-tailed test
*** $p < .001$, one-tailed test

tion—movement from minimal activity to fervent debate. Hence reinforcing much previous research, belief that one can influence politics appears to lead to greater participation in the process. Moreover, such feelings of political power appear to be subject to manipulation by message senders. Thus the path model appears to confirm that exposure to the value-priming stimulus increases individual participation through the mediating efficacy variable. But does this relationship show up when looking at it directly? In other words, do the data reveal a direct relationship between exposure to value priming and participation?

Column four of table 6.1 shows the degree to which exposure to the heresthetic stimulus directly influences activity in the simulated committee discussion. Exposure to the heresthetic stimulus accounts for a .55-unit increase in participation, everything else being equal. This relationship is statistically significant at the .05 level in a one-tailed test.[2] Listening to Limbaugh lecture on the salience of individual effort seems to spur listeners to action, independent of simple exposure to Limbaugh, conservative sentiment, gender, race, or other attitudes. Therefore while much scholarship has championed the role of efficacy as a determinant of participation, these findings go a step further, showing how efficacy can be used as a tool of mobilization. The results also provide additional support for heresthetic as an effective propaganda tool, expanding the framework to include persuasion as it applies to behavior inducement.

The following section goes beyond the lab to explore political participation more fully and investigates the constructionist model of talk-radio-induced participation. Specifically, I look at the degree to which talk radio may function like a social network, making those who agree with the political messages espoused on the Limbaugh show more efficacious and participatory, while making liberal listeners less efficacious and less participatory.

Constructing Reality from Pseudosocial Networks

Drawing upon the theory that talk radio audience members construct reality based on inferences drawn from the pseudosocial networks that talk radio creates, this section provides a more general assessment of the impact of listening to Rush Limbaugh on traditional measures of participation. I rely on data from the 1994–96 American National Election Study Panel (ANES). As mentioned in chapter 4, the sample includes 389 respondents who were part of a panel interviewed in 1994 and reinterviewed in 1995. I predict respondents' levels of participation in 1996 by their levels of Limbaugh listening in 1995, controlling for participation in 1994. In this section, the hypothesis differs somewhat from that tested in the previous section, because unlike the experimental stimuli, Limbaugh listening in the real world includes exposure to a steady stream of callers who reinforce the message of the host (see chapter 2). This presence of a perceived community or electronic town hall introduces a new dynamic. Whereas the simple priming of efficacy by Limbaugh may engender greater feelings of efficacy and therefore greater participation on the part of listeners, regardless of the ideological worldview of those listeners, the pseudosocial network of callers/listeners may encourage listeners to perceive that conservatism dominates the mind-sets of average Americans. Consequently, those who agree with the views expressed by the host and callers may be encouraged by the belief that others think the same way they do. On the other hand, listeners who disagree with the dominant views expressed by the "community" may come to feel isolated and discouraged.

Measurement

Political participation is "activity that is directly or indirectly aimed at influencing the selection of government personnel and/or policy outputs" (Verba and Nie 1972:2). So while many equate participation with voter turnout, participation also includes a variety of higher-level activities, including contacting public officials, working for a campaign or party, advising others about their vote choice, advertising for candidates, attending rallies or marches, and contributing financially to a campaign or party. Consequently, a complete measure of participation should include as many of these activities as data will allow.

To obtain as complete a measure of participation as possible, I followed a three-step process. First, I distinguished three related but distinct modes of participation: voting, proselytizing, and campaign activity. Second, when possible, I measured these modes of participation with indexes composed of multiple indicators. Third, after creating measures of each mode of participation, I placed each of these newly created measures in a factor analysis to confirm the unidimensionality of the underlying general concept of participation and to strip the latent concept of measurement error.

Voting, of course, is measured by whether the 1995 ANES pilot study respondents identify themselves as having voted in the 1996 presidential and midterm elections. *Proselytizing* involves trying to influence politics by persuading others to think and act in a particular manner regarding politics. Proselytizing is measured by two indicators: the extent to which people discuss politics, and whether people advise others about their vote choice. Twenty-one percent of respondents reported engaging in political discussion at least a few times a week and advising others' vote choices. *Campaign activity* involves attempting to influence political outcomes by (1) working for a party or candidate, (2) attending political rallies or meetings, (3) displaying political buttons or bumper stickers, and/or (4) contributing to a political party, candidate, or interest group. Sixteen percent of respondents reported engaging in activist behavior of some kind.

The Cronbach's alpha of the index summing these indicators of participation is .60, indicating an acceptable level of reliability scale. Similarly, factor analysis confirms the hypothesized single underlying dimension of participation, displaying an eigenvalue of 1.14. The latent factor loads on the variable measuring opinion leadership most heavily, followed by campaign activity, voting, and contacting House representatives. The working dependent variable measuring participation in the analysis is the factor score resulting from this factor analysis.

The independent variable of interest in this analysis is frequency of individual Limbaugh listening, measured in 1995. This variable is coded as "0" if a respondent does not listen to the Limbaugh program, "1" indicates that a respondent listens occasionally, "2" indicates that the respondent listens once or twice a week, "3" indicates that the respondent listens almost every day, and "4" indicates daily listening.

To account for the various other factors that regularly influence an individual's propensity to participate, and to ward off spuriousness, I have included the lagged measure of participation (1994) in the model as a control. This 1994 measure of participation is identical in construction to the 1996 measure and produced virtually identical reliability scores and factor loadings. Specified in this way, the model depicts the relationship between Limbaugh listening in 1995 and change in propensity to participate between 1994 and 1996. As discussed in chapter 4, panel analyses of this sort can greatly enhance causal inference because of the temporal sequence of the variables. Logically, change in one variable can only cause change in another variable if the former variable precedes the latter variable in time.

Results: Political Efficacy

If listening to Rush Limbaugh influences individual participation by the mechanism that I have proposed—that is, by priming the value of political efficacy—Limbaugh listening should produce significant changes in individual political efficacy as well as participation. Thus I have analyzed the relationship between Limbaugh listening and political efficacy for both (a) conservatives and moderates, and (b) liberals. I hypothesize that conservatives and moderates experience an enhanced sense of political efficacy as a result of listening to Limbaugh, because listening is akin to placing oneself in an environment where opinion is nearly consensual (conservative). Hence listeners come to feel vindicated and confident—ready to "spread the conservative gospel" to others, support conservative

Table 6.2 OLS Regression on Political Efficacy in 1996 on Limbaugh Listening and Lagged Efficacy

Independent Variables	Sample: Conservatives and Moderates	Sample: Liberals
Coefficient	Beta Coefficient	Beta
Limbaugh listening	.10**	-.01
Efficacy in 1994	.46***	.50***
Constant	NA	NA
N	239	113
Adjusted B^2	.24***	.23***

Source: American National Election Study Pilot, 1995

** *p* <05
*** *p* <01

candidates, and the like. On the other hand, liberals likely experience a diminished sense of efficacy, because their cherished beliefs are those ridiculed and satirized on a continual basis. Only the most self-assured liberal should be able to withstand such unremitting criticism and not come away somewhat more cynical about the process.

Table 6.2 displays the results of the relationship between Limbaugh listening in 1995 and political efficacy in 1996, controlling for lagged political efficacy (1994). Column 2 reports the results for moderates and conservatives. Listening appears to produce substantial change in political efficacy. The standardized regression coefficient of .10 means that a 4 standard deviation change, going from not listening at all to listening regularly, corresponds to an increase in participation by roughly 14 percentile ranks, a statistically significant relationship.

Column 3 of table 6.2 reports the relationship between listening and efficacy for liberals only. On this score, the data do not show a statistically significant relationship between increased listening and decreased efficacy. However, the sign is in the right direction, and the lack of statistical significance may be a result of bloated standard errors due to the small sample of liberals. To the extent that the null hypothesis of no relationship is correct, perhaps this is because many liberals who listen to

Limbaugh are those who are quite secure in their ideology, and are listening in order to "know thine enemy." Such listeners are not likely to be intimidated by even the most vitriolic criticism. The relationship might also be confounded by the presence of overt heresthetic in the Limbaugh message—priming listeners to feel as though they can make a difference in politics. Chapter 2 discusses how such polemics are not uncommon on Limbaugh's broadcasts, and the first section of this chapter demonstrates how such heresthetic may encourage greater efficacy even among liberal listeners. Thus the insignificant result displayed in the second column of table 6.2 may reflect social network dynamics and heresthetic dynamics working against each other.

Although listening does not appear to cause liberals to feel less efficacious, the data do show that listening produces more efficacy among moderates and conservatives. Therefore the theory on which this paper is based has survived a critical first test. But does this enhanced efficacy lead to more participation?

Results: Participation

Table 6.3 displays the results of the 1996 analysis of participation. The second column depicts the relationship among moderates and conservatives. Because participation is measured with a factor score, interpretation of unstandardized regression coefficients is difficult. But looking at the standardized coefficients reveals that a 1 standard deviation increase in listening produces a .14 standard deviation increase in participation, controlling for participation in 1994. In other words, knowing that a respondent listens to Limbaugh at least twice a week (a 2 standard deviation increase above the mean) corresponds to a mean increase in participation of 11 percentile ranks, a statistically significant relationship.

Unlike the examination of conservatives and moderates, I expected to see a substantial negative relationship between listening and participation when only liberals were included in the analysis. The third column of table 6.3 displays the relationship between Limbaugh attendance and participation, analyzing liberals only. A 1 standard deviation increase in Limbaugh listening produces a .09 standard deviation *decrease* in participation in 1996, controlling for participation in 1994. Hence movement along the listening scale from nonlistening to listening twice or more a week corresponds to an 8 percentile rank decrease in participation. This relationship does not pass muster in terms of statistical significance, so the reliability of the predicted coefficients is questionable. However, as

Table 6.3 OLS Panel Regression of Political Participation in 1996 on Limbaugh Listening, Including Moderates and Conservatives Only

Independent Variables	Sample: Conservatives and Moderates	Sample: Liberals
	Beta Coefficient	Beta Coefficient
Limbaugh listening	.14**	−.09
Participation—1994	.30**	.24***
Constant	−.10	NA
N	212	114
Adjusted R^2	.14***	.04***

Source: American National Election Study Panel, 1994–96

* p <10
** p <05
*** p <01

was stated earlier, the lack of statistical significance may be a function of bloated standard errors caused by a small sample.

Discussion

In sum, this chapter has examined the relationship between Limbaugh listening and political participation. I have applied and tested two separate theories of how participation levels of Limbaugh listeners might be affected by their listening proclivity. First, I extended the value heresthetic model described in chapter 3 to the realm of participation, positing that propagandists may mobilize audience members without making specific calls to action by simply speaking in such a way as to make audience members feel more efficacious. I tested this model with a legislative committee experiment. I found that when subjects had been randomly exposed to a message designed to prime political efficacy, they were not only more likely to report greater feelings of political efficacy, but were also substantially more likely to engage in spirited argument during deliberations over the passage of a hypothetical social welfare spending bill. These findings support the notion that not only do psychological constructs such as political efficacy significantly determine participation, but that such efficacy is not necessarily exogenous—it can be affected at least in the short run by media messages.

Second, I tested whether the talk radio *medium*, rather than the message itself, could have an effect on listener efficacy and participation by encouraging listeners to construct a vision of political reality that includes disproportionate strength of conservative political opinion among "ordinary" citizens because of the dominance of conservatism in the messages expressed by callers. By encouraging such constructed reality, the call-in format may induce social network effects, where listeners who agree with the views expressed on the show come to feel emboldened, while those who hold different views are discouraged. Using a panel design to assess attitude change, I found that the more that moderates and conservatives listened to the *Rush Limbaugh Show* in 1995, the more politically efficacious they became over time—becoming more participatory in politics in 1996 than they had been in 1994, a year in which many pundits attributed the strong Republican success to the rise in talk radio listening. While the substance and sign of the coefficients indicated that liberal listeners became less efficacious and participatory over time, the relationships were not statistically significant. But the sample of liberal listeners was very small ($n = 62$), surely reducing the efficiency of the estimates. Further research should be conducted with larger samples to reassess the degree to which listening discourages activity on the part of liberals. In general, these findings support the notion that individuals' decisions about whether and how much to participate in politics are affected by the social networks in which they travel. These findings also imply that the concept of a social network transcends the traditional notions of family, friends, and community organizations, and includes nontraditional electronic networks such as a talk radio audience.

This chapter also serves as something of a segue, because it extends the heresthetic model of persuasion as it may apply to talk radio, while introducing the constructionist model of belief change as applied to talk radio. The following chapter continues to apply constructionist theory to the study of talk radio effects over audience members. Moving away from the study of what audience members choose or do, chapter 7 analyzes what people believe as fact, testing the hypothesis that the information and misinformation levels of individuals are a function of constructed realities inferred from often implicit media messages. Chapter 7 also differs from the previous chapters in that it expands the measure of talk radio listening to encompass more than just the *Rush Limbaugh Show*.

7 Information, Misinformation, and Political Talk Radio

C. Richard Hofstetter and David C. Barker with James T. Smith, Gina M. Zari, and Thomas A. Ingrassia

One of the theoretical foundations for democracy is an informed citizenry. Much research has considered the extent to which the American electorate possesses the requisite sophistication to execute republican government (Converse 1964; Nie, Verba, and Petrocik 1979; Smith 1989). Many have concluded that although the majority of Americans may not be terribly informed, the uninformed take cues from the smaller percentage of sophisticated "opinion leaders" (Katz and Lazarsfeld 1964; Zaller 1992). However, fewer scholars have considered the ramifications of a *misinformed* citizenry. Misinformation, or *erroneous understanding*, differs dramatically from simple ignorance, or *the lack of understanding*. Often, the misinformed may even hold their incorrect beliefs *with confidence* (Kuklinski et al. 1997). Hence the difference between the uninformed and the misinformed may be akin to the difference between staying home on election day versus holding a placard at a rally. The uninformed are likely to opt out of politics, or to rely on heuristic measures such as party identification or opinion leaders, thus enabling them to potentially behave as if they were informed (Page and Shapiro 1992; but see Bartels 1996). By contrast, the misinformed may participate at high levels—writing to Congress, proselytizing for a candidate, contributing money, and so on. As such, although an uninformed citizenry might not pose a great threat

Much of this chapter has been reprinted with permission from Hofstetter and Barker (1999).

for democracy, the presence of a largely *misinformed citizenry* may misdirect electoral outcomes and even policy direction.

Media often serve as the democratic marketplaces in which ideas are marketed, bought, and sold. Ideas presumably compete on the basis of merit and value, and democracy succeeds when the collective public rationally chooses its preferred idea, votes accordingly, and watches the preferred policies ensue (Page and Shapiro 1983). During the last century, mass media have come to relish this role, and have largely moved away from the "partisan press" to more "objective" journalism (but see Hallin 1995). However, some have compared political talk radio to the partisan press of the nineteenth century, where ink has been exchanged for airwaves (Jamieson, Capella, and Turow 1996).

How does misinformation spread? Misinformation may not always result from the dissemination of false assertions. Often, individuals may draw false conclusions by making grand inferences from bits of incomplete information. When considering a political issue about which we have only partial information, we may "fill in" the missing pieces with contrived information that matches our established worldview (Shneidman 1969; Kuklinski et al. 1997). This *inferential reasoning* theory of political "learning" guides our present analysis of talk radio and information/misinformation. We explore the extent to which regular and active listening to political talk radio may lead to greater levels of both information and misinformation.

Because of the engaging, often volatile, somewhat ambiguous, and always repetitive character of talk radio content, we hypothesize that listening may increase levels of information regarding nonideologically charged matters (such as which party controls the House of Representatives) while simultaneously corresponding to misinformation regarding ideologically charged affairs (such as whether the deficit increased under the Clinton administration and whether America spends more money on welfare than defense). Regarding misinformation, we posit that listeners may be induced to draw conclusions that are counterfactual but are nevertheless consistent with the general ideological tenor of the medium.

Beyond simple exposure to the medium, attitudes and behaviors may be a consequence of personal involvement with the medium, with construction of social and political realities driven by interaction between listeners and political talk programming and perceptions of the medium (Schoemaker, Schooler, and Danielson 1989; Newhagen 1996; Hofstetter 1996; Hofstetter and Gianos 1997). Active communication is associated with enhanced media influence similar to the processes involved in traditional news media (Chaffee and Scheudler 1986; Grunig 1989).

Most important for purposes of this study, exposure to and involvement with political talk radio has often been associated with relatively high levels of political information (Bolce, de Maio, and Muzzio 1996; Hofstetter 1996; Hollander 1995), albeit not universally (Weaver 1996: 42). Despite the flamboyance of many hosts and messages, audiences nonetheless appear to hold higher levels of information in association with involvement with political talk. Specific mechanisms of priming, agenda setting, and episodic sensitization to politics may account for information gain among otherwise casual, sometimes politically inert, listeners (Hofstetter 1996).

Despite enhancing the level of political information among audiences, political talk radio may also enhance political misinformation among the same groups. In one of the few careful studies of political talk content, Davis and Owen (1998) found that shows are orchestrated to maximize audience engagement and entertainment rather than to serve as a public forum. Hosts rarely make bold counterfactual assertions, but more usually populate programming with the invective of sarcasm, innuendo, and diatribe, repeatedly directed against targets viewed as "liberals," liberals' policy preferences, the president, and Democratic activists. Such content may lead listeners to draw conclusions from programming as information is processed, especially when information is otherwise lacking and the general thrust of program content appears to define reality, or when predispositions to draw specific conclusions are already present and listeners are psychologically "ready" to make an inference (Shneidman 1969).

In a highly innovative study, Kuklinski and colleagues (1997) found misinformation about welfare policy to be widespread, correlated with conservative views on social welfare, and relatively impervious to change except when confronted directly and bluntly by irrefutable information. Consistent with Shneidman's observations about the psychology of inferential reasoning, Kuklinski and colleagues argued that people tend to "fill in" missing portions of political schema and stereotypes, usually in ways that are consistent with other beliefs they harbor.

We believe that political talk radio may engender such inferential reasoning on the part of its listeners. Talk radio likely stimulates attention among audiences and engages listeners (Hofstetter 1998:283–285). The more attentive and engaged are likely to become more aware of diverse aspects of political problems through the processes of priming, content sampling, and latent learning (Zaller 1992:333–337). It is a short step from awareness of such program content diversity to inferences that generally coincide with existing personal biases or the general direction of arguments apparent in programming. Bits of information are acquired; some are correct and others are false. Once interest in issues is engaged and relevance is demonstrated by talk hosts, audiences may come to draw inferences about candidates, issues, and events under the formative influence of the programs. The following section describes the procedures we used to evaluate this theory.

Research Design and Methodology

The Sample

Data were drawn from a random-digit-dial telephone survey of 882 English-speaking adults (persons eighteen or older) living in households that could be reached by residential telephone in the San Diego metropolitan area (more than 96% of all households).[1] Exposure to political talk radio was measured by asking participants: "We are interested in how often people listen to political talk shows on the radio, that is, radio programs that are about politics and government, where people call and talk to the host. About how many times during the last month have you listened to (Limbaugh, Matalin, Liddy, Leykus, Hedgecock, Suarez, some other talk show) on radio programs that are primarily about politics or public affairs?" Respondents who had not listened during the last month were asked if they had "ever listened" to political talk. This summated scale measured simple exposure to political talk. Thus separate measures of listening were created for Limbaugh, Liddy, Matalin, Hedgecock, Leykus, NPR/Suarez, and "other." Each variable measuring exposure to an individual host ranged from 0 to whatever the highest number of listens had been in the previous month—for example, 31 for Hedgecock, 30 for Limbaugh, 15 for Matalin, and 12 for Liddy (although the shows only broadcast live five days a week, reruns air on weekends for Hedgecock and Limbaugh).

From these individual scales, two indexes of political talk radio exposure (conservative talk radio listening and moderate talk radio listening) were computed using a principal components analysis of the political talk radio exposure measures, rotated to simple structure by varimax criteria.

Exposure to Hedgecock, Limbaugh, Matalin, and Liddy (rotated loadings of .80, .78, .67, and .47, respectively) loaded on one factor, while exposure to "other hosts," Suarez/NPR and Leykus (rotated loadings of .76, .71, and .59, respectively) loaded on the other factor. The first factor, labeled "conservative hosts," explained 28.4 percent of the total variance in the exposure data, while the second factor, labeled "moderate hosts," explained 22.1 percent of the total variance.[2]

Political Talk Activity Beyond measuring simple frequency of exposure, we also constructed a scale of political talk *activity* (Hofstetter 1998). The scale includes respondents who do not listen, those who just listen, listeners who talk to someone else because of something heard on a show, listeners who both talk and take action because of something on a show, and listeners who talk, take action, and call talk shows. The political talk radio activity scale was unidimensional and internally consistent, with a coefficient of reproducibility exceeding .96.

Table 7.1 describes the mean differences between nonlisteners (58% of the sample), those who just listen (12%), those who listen and talk (20%), those who listen, talk, and act (9%), and those who listen, talk, act, and call (just more than 2%) with respect to political interest, party identification, and ideology. As the table shows, political talk activity

Table 7.1 Political Talk Radio Activity Scale Types by Selected Variables

PTR Activity	Political Interest	Party	Ideology	Percent
Nonlisteners	2.66	4.34	3.09	57.5
Listeners	2.83	3.48	2.70	11.6
Listen/talk	3.03	3.68	2.72	20.0
Listen/talk/act	3.32	3.41	2.72	8.5
Listen/talk/act/call	3.56	3.12	2.41	2.4
F	19.81	7.18	6.20	100.1
Df	4,745	4,665	4,657	(764)
P<	.001	.001	.001	

Statistics were based on the 764 responding to the items. Numbers in the cells are means of variables in the columns. Statistics for one-way analysis of variance are reported below the means. The percentage of respondents in each political talk radio activity scale type is presented in the right column. High means indicate greater misinformation, information, political interest, more-Democratic identification, and more-liberal identification.

correlates with political interest. Such a finding stands to reason, because only the politically interested would be inclined to listen to talk radio, talk to others about politics, take action, or call a show. Furthermore, not only do conservatism and Republican Party identification increase with more exposure to talk radio (as we noted previously)—these things also increase as one becomes more actively engaged with the medium.

Measurement of Dependent Variables

Political information was measured following Delli Carpini and Keeter (1996) and was conceptualized as holding correct beliefs about the political world. A conventional political information scale was formed by summing the responses to a series of questions about the structure and process of American government. Items included knowledge of who determines if a law is constitutional (Supreme Court), who nominates judges to federal courts (president), the size of a majority required to override a presidential veto (two-thirds), the party with the most members in the House of Representatives (Republicans), which party is more conservative (Republicans), the length of term for a U.S. Senator (six years) and for a U.S. House member (two years), and who is the current Vice President (Al Gore). The scale mean was 6.3 (SD = 1.7) and attained an adequate, if not high, level of reliability (Kuder-Richardson 20 = .66). Wording and distributions of all statements used to compile the misinformation index are displayed in table 7.2.

Misinformation Measurement of misinformation involved several steps. First, the program content of the most popular nationally syndicated shows (Limbaugh, Leykus, Matalin, Suarez, Liddy) and local political talk hosts (Hedgecock) that broadcast into the San Diego media market was sampled and taped during a one-month period (every third program) immediately prior to the survey. To determine the extent to which listening engenders inferential reasoning, coders were instructed to write down every statement that occurred to them spontaneously while they listened to programming. Programming was taped so that content could be reviewed and additional statements written. The process resulted in more than three hundred statements. The statements were edited and condensed so that they constituted simple declarative assertions.

Thirty-two divergent assertions that could be documented as incorrect by easily accessible public information sources (primarily mass me-

Table 7.2 Information Scale Items and Distribution

	Percentage		
Statements	Incorrect	Correct	DK
Whose responsibility is it to determine if a law is constitutional or not?	21	75	4
Whose responsibility is it to nominate judges to federal courts?	27	59	14
How much of a majority is required for the U.S. Senate and House to override a presidential veto?	22	59	19
Do you happen to know which party has the most members in the House of Representatives in Washington?	12	74	13
Which party, if any, is more conservative?	13	79	8
Can you tell me the length of term for a U.S. Senator?	42	43	15
Can you tell me the length of term for a U.S. House member?	42	37	21
Can you tell me the name of the current Vice President of the United States?	3	89	8

Respondents were told: "Now we would like to ask you a few questions about the way the political system in this country works." Correct scores were coded as 1. The information scale was constructed by summing the 1 responses to individual items. The scale Mean = 6.3, SD = 1.7, and Kuder-Richardson 20 = .66.

dia) and that were judged to be "reasonable conclusions" from program content by investigators, were selected for inclusion in the misinformation scale. Pretesting resulted in the elimination of fourteen statements. Wording and distributions of all statements used to compile the misinformation index are displayed in table 7.3.

Finally, survey participants were read the list of incorrect statements, and were asked to evaluate each statement in terms of its veracity. The misinformation scale was constructed by summing the number of times participants indicated that they were sure that statements were true or thought that they were true, indicating that participants had some confidence that the incorrect statements were accurate. The resulting scale was highly reliable (Kuder-Richardson 20 = .82, with scores ranging

Table 7.3 Misinformation Scale Items and Distributions

Statements	Percentage		
	Incorrect	Correct	DK
a. Most people are on welfare because they do not want to work.	43.5	52.3	4.2
b. Illegal immigrants get most of the jobs in this area.	34.2	59.3	6.5
c. Illegal immigrants commit most of the crimes in this area.	20.4	72.7	6.9
d. Test scores in public schools have dropped sharply in the last 20 years.	66.0	23.6	10.4
e. Pregnancy by unwed teenagers continues to increase rapidly.	71.7	22.8	5.4
f. Bill Clinton has been indicted for illegal activities in Arkansas.	29.0	55.9	15.1
g. Hillary Clinton was found to have been implicated in Vince Foster's death in Washington.	19.6	59.0	21.4
h. Growth in the budget deficit has increased during the Clinton presidency.	38.8	48.8	13.2
i. Unemployment has increased during the Clinton presidency.	21.4	69.3	9.4
j. America spends more on foreign aid than on law enforcement.	61.1	26.0	12.8
k. America spends more on welfare than on defense.	35.4	52.5	12.1
l. President Reagan cut the national deficit.	25.2	62.0	12.8
m. Teaching about religious observations is illegal in public schools.	56.2	36.8	7.0
n. Giving clean needles to drug addicts has increased AIDS in California.	14.2	73.5	12.2
o. Most of the homeless in America are too lazy to work.	26.2	67.8	6.0
p. Nearly all Americans oppose sex education in public schools.	18.6	74.6	6.8
q. Most Americans are opposed to abortion.	24.6	66.8	8.6
r. More taxpayer money is spent for abortion than on care for the elderly.	25.1	57.4	17.5

Respondents were asked: "Following is a list of things that some people think are true and others think are untrue. For each in the following list tell me whether you are sure that it is true, think that it might be true, or are sure that it is untrue." The misinformation scale was constructed by summing "sure" that the statement is true and "think that it might be true." Mean = 10.1, SD = 4.8, and Kuder-Richardson 20 = .82 for the scale.

from 0 to 22, mean = 10.1, and SD = 4.8). "DK/NA" (don't know/no answer) responses were not counted as misinformation, because the response indicates lack of information rather than the presence of erroneous information.

Intercorrelations and Model Specification

The major hypotheses of this study are that exposure to political talk radio increases both information and misinformation. Before testing the hypotheses, it is useful to explore several correlates of the major scales used to provide some evidence of the validity of the scales. The Pearson's correlation (r) between political information and misinformation scales was, as expected, −.20 ($p < .001$). The moderate correlation suggests that although the two scales are related, information and misinformation are not simply opposite ends of the same continuum. Education was correlated .35 ($p < .001$) with information and −.15 ($p < .001$) with misinformation, also as expected. However, while political interest is correlated .33 ($p < .001$) with information, it is not significantly correlated with misinformation.

Most of the political talk content that listeners report in the San Diego market is conservative in tone. Statements in the misinformation scale endorse beliefs of the kind that one might expect to be reflected in this programming. Thus correlations in the sample between misinformation and partisanship (−.14, $p < .001$) and ideology (−.16, $p < .001$) are expected. The political information scale was also correlated with partisanship (−.10, $p < .008$), but not with ideology. Income and age were correlated with information (.26, $p < .001$, and .11, $p < .001$, respectively) but not with misinformation.

With respect to the intercorrelation among different measures of talk radio listening used in this study, we expected the three items to share some variance, especially because political talk activity is derived in part from the exposure scales. As expected, moderate talk radio listening is correlated with conservative talk radio listening (.26, $p < .001$), and political talk activity is correlated with both moderate talk radio listening (.28, $p < .001$), and with conservative talk radio listening (.26, $p < .001$). These correlations are not strong enough to pose a serious threat of multi-colinearity in our models, and thus we choose to simultaneously include them in all of our equations. However, the significance of the correlations does indicate that the variables compete for variance, making the statistical associations that we observe in our samples somewhat conservative estimates of the true relationships.

In an effort to ward off spuriousness in our models, we controlled for race, sex, party identification, ideology, age, income, education, and political interest.[3] As expected, political talk activity is significantly correlated with political interest (.28, $p < .001$) and education (.11, $p < .04$). Conservative talk radio listening is significantly correlated with age (.19, $p < .001$), interest (.17, $p < .001$), income (.10, $p < .03$), Republican Party identification (.24, $p < .001$) and conservative ideology (.21, $p < .001$). Moderate talk radio listening was significantly correlated with political interest (.18, $p < .001$), age (.08, $p < .04$), and (somewhat surprisingly) Republicanism (.09, $p < .03$), but not conservatism. These significant correlations between the independent variables of interest and the controls confirm the earlier results obtained by examining differences in means for these variables between nonlisteners, listeners, and active listeners. More important, these correlations provide evidence for the necessary inclusion of these controls in our models.

Findings

Political Talk Radio and Information

As the first four columns in table 7.4 display, political talk radio activity independently and significantly corresponded to political information ($p < .01$). Substantively, a respondent who listened, called, talked, and took action because of something that he or she heard on talk radio tended to, on average, know the answer to one more item on the information scale than did nonlisteners, a difference of 12.5 percent. It is interesting to note that frequency of exposure to conservative talk radio displays a significant negative correlation with political information, indicating that although conservative talk radio listeners are more interested in politics, read the newspaper more often, and are more likely to vote, they are less likely to hold accurate beliefs even regarding nonideological facts (such as which branch of government determines the constitutionality of a law) when other items are controlled, such as political talk activity.[4] Moderate talk radio listeners also appear to be somewhat less informed, but this relationship is not statistically significant. The model explained about 22 percent of the variance in information ($R = .47$, $F_{(9,755)} = 23.27$, $p < .001$).

Table 7.4 Regression of Information and Misinformation on Talk Radio Exposure and Activity

	Information			Misinformation		
Predictors	*B*	std.e.	*P*<	*B*	std.e.	*P*<
Political interest	.68	.10	.001	−.05	.18	.382
Age	.00	.00	.173	−.01	.01	.162
Education	.22	.03	.001	−.20	.06	.001
Income	.12	.04	.001	−.06	.07	.179
Partisanship	−.08	.04	.030	−.08	.08	.142
Ideology	.07	.08	.159	−.28	.15	.035
Conservative PTR	−.15	.09	.049	.37	.16	.010
PTR activity	.26	.09	.002	.25	.16	.059
Moderate PTR	−.11	.08	.098	−.33	.16	.017
(Constant)	−.95	.54	.048	10.96	1.00	.001
R	.47			.25		
$F_{(9,755)}$	24.26			5.53		
P	.001			.001		

Note: Numbers are unstandardized regression coefficients, standard errors, and associated one-tailed probabilities for the regression of the political information and misinformation scales on the selected predictor variables. Regressions were computed using substitutions of means for missing data, assuming that data were missing at random. Where missing data were deleted pairwise, however, the minimal *N* was 765 for the analysis. We replicated this analysis to assign the political information scale to the list of independent variables. The substance of the findings remained identical to that of the original analysis, even with political information being the most powerful negative partial predictor of misinformation in the equation. The presence of political information as a predictor increased the R^2 to .32.

Political Talk Radio and Misinformation

As the last three columns in table 7.4 attest, the regression of misinformation on the same predictors resulted in different associations. The overall regression was statistically significant ($R = .26$, $F_{(9,755)} = 6.19$, $p < .001$). Partial associations between education and ideology and misinformation were statistically significant, while those for political interest, age, income, and partisanship were not. Most important for purposes of the analysis, the partial association between political talk activity and misinformation was not statistically significant (although in the direction predicted by the hypothesis). However, as hypothesized, listening to conservative talk radio was positively associated with misinformation ($B = .37$, $p < .01$) regarding ideologically charged facts.

Thus it appears that not only are conservative talk radio listeners in the sample less informed about general information than nonlisteners, the conservative talk devotees tend to be more misinformed as well, likely drawing false inferences from show content about political facts (such as whether the deficit has increased or decreased under President Clinton, and whether he has been indicted for illegal activities in Arkansas).

Somewhat surprisingly, the more one listens to moderate talk, the *less* misinformed one tends to be regarding these matters, even though moderate talk devotees tend to be more Republican than their nonlistening counterparts. Perhaps hearing neutral or positive references to Democrats, liberals, and the Clinton administration leads listeners to draw inferences that, in this case, happen to be true. Of course, these findings do not mean that moderate talk radio programming necessarily does a better job than conservative talk at providing listeners with accurate information. Those inclined to listen to moderate programming may be more fair minded than conservative talk listeners, something that the shows themselves cannot control. Even more likely is that these moderate shows may at times lead listeners to draw false conclusions regarding assertions that fit a liberal mind set. In this paper, we do not explore that possibility.[5]

Discussion

Analysis of data provided partial support for hypotheses in this study. Personal involvement with political talk radio, indicated by a political talk radio activity scale, was associated with increased information, controlling for amount of exposure to conservative and moderate political talk shows and for a series of other predictors, including partisanship, ideology, political interest, education, age, and income. This finding is consistent with prior studies of the medium (Hofstetter et al. 1997). Interest, education, income, and (Republican) partisanship were also associated with increased information.

More important, exposure to conservative political talk shows was related to increased misinformation, while exposure to moderate political talk shows was related to decreased levels of political misinformation, after controlling for other variables. Partisanship and ideology were particularly important statistical controls, because the content of much political talk programming is critical of Democratic political leaders and what are labeled as liberal policies.

Thus the analyses suggest that involvement with political talk is associated with enhanced political information among listeners, but that exposure to more-conservative hosts also may increase misinformation regarding ideologically charged matters. On the other hand, exposure to moderate hosts (including NPR hosts) may decrease misinformation. We suspect that the association between moderate talk radio listening and lower levels of misinformation can be explained by the content of the shows and the construction of our misinformation scale. Our misinformation scale is primarily made up of items that ideological conservatives are likely to embrace. Therefore conservatives who listen to conservative talk radio are more likely to infer from the program content that these statements are true than are moderates who listen to moderate political talk. We suspect that the inclusion of more liberally charged misinformation items might wash out some of the negative association between moderate talk radio listening and misinformation. Future research should test whether listening to moderate and liberal talk shows corresponds to greater misinformation regarding liberally charged assertions.

These findings support the theory that individuals rely on inferential reasoning when trying to assimilate political information. Individuals appear to construct their own political reality by extrapolating from the incomplete bits of information that are available to them and fitting that information into schemata that are already structured around a particular ideological worldview. This inferential reasoning appears to often lead individuals to hold incorrect beliefs with some degree of confidence. Such widespread misinformation based on inferences from talk radio content alters our impression of democracy as a forum of competing ideas, with media serving as the marketplace where those ideas compete. Perhaps the growth of talk radio is contributing to a change in the operation of democracy in the United States. Just as supermarkets (and now hypermarkets) replaced neighborhood corner markets, perhaps our "marketplace of ideas" now functions more like a "supermarket of ideas," where ideas compete not so much on the basis of merit but on the basis of flashy advertising and window dressing.

While the findings of this community study are limited in time and space, and by difficulties associated with cross-sectional studies—especially difficulties in drawing inferences of causality—the conclusions are sufficiently provocative to encourage additional research regarding both the power of inferential reasoning as a theory for political information processing, and the political impact of talk radio.[6]

8 Conclusion

This book has examined what might be termed the DNA of democratic politics—persuasion. Essentially, I have sought to understand how this democratic DNA is formed. In mapping this political genome, I have analyzed political behavior in its various forms—attitudes, candidate appraisals, policy preferences, partisan attachments, value orientations, vote choices, participation decisions, and belief structures. Each of these behavioral elements involves some kind of judgment. A democratic citizen is constantly engaged, asking herself, "Does that idea make sense? Which candidate do I prefer? What do these parties stand for? What is more important to me? Should I vote in this election? Can I make a difference? Who is depending on me? Is that true?" Is it possible that the dynamics of such American political judgment are increasingly becoming induced, simplified, or otherwise *Rushed* in the face of the information revolution that has spawned various forms of "new media?"

I suggest that the answer is an unqualified "yes." The results obtained from the various analyses described in this book, which have focused on one of the most conspicuous forms of the new media—call-in political talk radio—speak to the way we understand the social psychology of modern political judgment and the changing role of mass media as it relates to modern democratic decision making.

Although talk radio is an important medium that shows no signs of withering away, it is but one salient example of the new ways that Americans inform themselves about public life. To summarize, more choices and shortened attention spans have led Americans to combine activities wherever possible. Now we want to be entertained while we are being informed, and we do not want to wait around for it. This demand has led to the explosion of television news magazines such as Nightline, cable political programming based on bombast (i.e., *The Spin Room, The O'Reilly Factor, Crossfire*), comedic political programming *(The Daily Show, Politically Incorrect, That's My Bush),* and Internet sites devoted to particularized tastes *(Salon.com, The Drudge Report).* So this book speaks to more than the influence of a single media personality or a single medium, but instead seeks to serve as a gateway to exploring the consequences, in terms of political communication, of the "information revolution."

This chapter will review the major theoretical and applied political questions addressed in the preceding chapters, summarize the results, and discuss the possible implications of these findings.

Understanding Political Persuasion

As I have argued throughout this book, persuasion is the foundation of democratic politics, where power cannot be achieved, organized, or maintained via simple coercion. Therefore the primary theoretical puzzle addressed in this book has been how political persuasion occurs—how can a propagandist further his interests by convincing audience members to make choices that they would not necessarily otherwise make? Social psychologists have taught us that there are two primary routes to persuasion—a central route and a peripheral route. And the role of heuristics (or peripheral-route decision making) has enjoyed a considerable amount of attention from political scientists over the course of the last decade. But a paucity of political research has tried to apply heuristic decision making to political persuasion directly. And very little attention has been paid to understanding the relative utility of central- versus peripheral-route strategies of persuasion under different conditions. Borrowing from Riker's distinction between heresthetical and rhetorical arts of political manipulation (1986), I have argued that rhetoric is the attempt to persuade via the central route—convincing an audience member to change his or her mind by presenting new, compelling information previously unconsidered, but within a single consideration "frame." Hence the audience member becomes convinced by something newly learned—either rationally or emotionally. On the other hand, heresthetic utilizes the peripheral route to persuasion, manipulating the relative salience of considerations upon which decisions are to be made through a joint process of framing and priming. Thus I explicitly conjoin framing and priming theory (while recognizing their conceptual distinctions) under the single rubric of heresthetic.

Previous analyses of political persuasion have bemoaned the efficacy of persuasive (e.g., campaign) efforts because they have assumed the route to persuasion is exclusively central. To elaborate, researchers have noted that those who are the most sophisticated political observers are the ones most likely to absorb persuasion attempts but are the least likely to accept the message and adopt a new point of view because, armed with considerable knowledge, there is relatively little new information to be absorbed, they selectively accept new information anyway, and they know how to counterargue against propaganda with which they disagree. But such assumptions fail to consider peripheral, or heresthetical routes to persuasion, whereby the sophistication of audiences is used to the propagandist's advantage. The propagandist simply frames a message in such a way so as to prime potentially important considerations (that work to the advantage of the propagandist) to the front of the audience member's head. Such priming will be more effective when audience members are sophisticated enough so as to already possess well-developed stores of considerations. Moreover, there is no reason for sophisticated audience members to apply their well-honed skills of counterargument, because the propagandist is not asking them to accept or believe anything that they do not already accept or believe—the propagandist is merely slyly asking them to apply one set of considerations to a question rather than another set, thus setting the cognitive agenda for the audience.

In chapter 3, we applied the first direct empirical test of the relative utility of heresthetic versus rhetoric as an agent of persuasion in the realm of politics. Focusing on relatively sophisticated audience members who were predisposed to be hostile to the message of Rush Limbaugh (thus minimizing the potential for persuasion), we found that even seven minutes of exposure to heresthetic made audience members five times as likely to adopt Limbaugh's position regarding governmental efforts to assist the poor. This effect for heresthetic is made more impressive by the fact that rhetoric—or the traditional effort to persuade via the central route by convincing an audience member that some outcome is less desirable and more likely if policy A is adopted—failed to produce any attitude change whatsoever, with audience members displaying identical posttest attitudes to those who had been members of the control or "placebo" group. These results provide the first contemporary, scientific empirical support for the formal models predicting the same results, which were first developed by Riker (1986, 1990), and had previously been supported by historical case studies (Riker, 1996).

Further, we contend in chapter 3 that the apparent success of heresthetic as a means of winning over an audience may be attributable to the frequent reliance by Limbaugh on core democratic values as the considerations being primed. Limbaugh deftly frames issues around the principles of freedom, self-reliance, and personal responsibility—values that form the core of the American ethos (e.g., Feldman 1988). Much research has determined the power of core values as choice determinants, but little research has attempted to unearth the dynamics of this process. Given the place of values as *core* principles, stable over time, few have considered how such base-value structures may be manipulated. We have not argued that Americans' base level of support for particular values such as freedom can be easily manipulated by a propagandist—such an attempt would be an example of rhetoric. Rather, we argue that propagandists may take advantage of the value pluralism among the American public—whereby sets of conflicting values are simultaneously cherished by virtually everyone—to prime one set of values to be more salient than another. This involves little risk on the part of the propagandist, because almost no one socialized in the United States counterargues with the notion that freedom (for example) is good and should be encouraged through public policy.

Thus I have attempted to bring together previously disparate strands of political psychology research. One strand has focused on the importance of core values as evidence that Americans are sophisticated, evolved political animals who look beyond their noses to determine not only "what's in it for me" but also "what is morally right"; use such principles to structure attitudes; and make relevant choices accordingly. The other strand focuses on the nature of framing and priming effects, noting that Americans seem to be ambivalent and easily manipulated into making particular choices based on which considerations are most salient in their minds at the moment. I suggest that both research streams are correct. But how can citizens be sophisticated philosophical observers who develop core principles and meticulously apply them to particular problems *and* allow themselves to be so easily manipulated by the ways that opinion leaders phrase questions or edit news stories? The answer may lie in the fact that Americans care very deeply about certain core principles or values—freedom, equality, self-reliance, personal responsibility, community, humanity, tolerance, justice, and so on, which, if taken to their logical extremes, often imply opposite policy outcomes. But through our political socialization, we have been taught to not experience cognitive dissonance over holding on to such potentially discrepant values. As such, we are not only value pluralistic but also value ambivalent—we want to maximize freedom *and* equality. But by calling one value *yin* (say, freedom) to the front of the head at the expense of its corresponding *yang* (equality), a citizen may temporarily fail to consider the unprimed value as heavily as she does the primed value. Then, when the citizen goes through the subconscious process of allowing values to structure policy preference and vote choice, she will do so primarily based on the primed value at the expense of the unprimed value.

In sum, the theoretical implications of the analyses undertaken in this book, as they relate to the social psychology of persuasion, speak to the importance of both hierarchical value models of political cognition and priming models of opinion expression. I have argued that such models are not mutually exclusive but in fact complementary, and that the joint process serves to enhance the likelihood of political persuasion when audience members are relatively sophisticated, compared to what can be achieved by the propagandist employing traditional rhetorical tools.

Deliberative Democracy

Part of the reason that persuasion is so central to democratic politics involves the theoretical ideal of a deliberative citizenry engaged in public discourse, debating the relative merits of ideas in an open "marketplace." Over the last ten years in particular, normative theorists as well as empirical political scientists have reignited an interest in the etiology and consequences of such deliberation (e.g., Fishkin 1995; Page 1996). The study of talk radio is directly relevant to this discussion because of the potential for an "open forum" where citizens can debate the merits of competing philosophies, thus creating a pseudo town hall. It is thought that such deliberation produces more thoughtful policy choices and perhaps an enhanced sense of community on the part of the participants. Such civic-mindedness and personal empowerment is often credited as the primary source of political participation on which our democracy rests (e.g., Putnam 1993).

In chapter 6, I tested the hypothesis that the talk radio airwaves serve as a pseudocommunity for listeners, encouraging political participation on the part of listeners (at least like-minded ones). The argument has two parts: First, I argued that the sense of personal empowerment, or internal political efficacy, on which political engagement rests (e.g., Verba and Nie 1972), can be primed just like other political values, and that such efficacy priming indirectly leads to greater political engagement. Using controlled experiments where one set of subjects was exposed to a message designed to make subjects feel like they "could make a difference," compared to a control group that did not receive such priming, I found that those exposed to the efficacy-priming stimulus not only reported feelings of much greater internal efficacy in posttest survey responses, but also were significantly more likely to participate (often intensely) in simulated legislative committee deliberation over the merits of a hypothetical social spending bill. As such, this analysis provides support for the notion that efficacy is not static, but can be prompted with relatively little effort on the part of an opinion leader. Moreover, persuasion in the form of mobilization can be achieved on the part of a propagandist by the use of heresthetic in the form of efficacy priming, without any direct appeal or call to action.

The second prong of the community-based argument has to do with the medium of call-in talk radio itself. I hypothesized that listening to a constant stream of fellow citizens argue in a consistent ideological direction encourages listeners to feel as if they are part of a larger "ideological community," which may lead to an enhanced sense of responsibility to that community as well as a general feeling of power due to the perception that they are part of something larger than themselves. But I argued that this dynamic should only occur for those who are at least open to the message being consistently propagated. For that minority of citizens who consistently receive the message but actively oppose it, I contend that the opposite may occur: these listeners may begin to feel isolated and unsure of themselves, perhaps even threatened within the perceived social community. The practical result, I argue, is that such listeners may participate *less* as a result of "living within the pseudocommunity."

I tested these hypotheses by breaking up regular Limbaugh listeners into two camps: (1) conservatives and moderates, and (2) liberals (a smaller, but not insignificant group). I found that as conservatives and moderates listen to Limbaugh over time, they become increasingly likely to express feelings of personal efficacy and become more likely to participate in the political process, in the form of voting, proselytizing, calling Congress, joining or working for the Republican Party, contributing to campaigns, and so on. Correspondingly, I found that liberal listeners become less efficacious over time and less likely to participate in politics—holding everything else constant. While both effects (for the conservatives and the liberals) were meaningful in a substantive sense, the relationship between listening and activity among the liberals did not quite achieve statistical significance, meaning that I do not have as much confidence in those findings as I do in those involving the conservative and moderate listeners. Nevertheless, the findings are supportive of Noelle-Neumann's (1984) classic hypothesis that when people are surrounded by others of vocally dissimilar mind, they may fall into a "spiral of silence," whereby political participation (especially public participation) is stifled.

Finally, with regard to deliberative democracy, my colleagues and I sought to understand the effects that call-in political talk radio might have on public levels of information and misinformation. Does the deliberation afforded by the medium lead to a more thoughtful and informed citizenry, or to more demagoguery? We looked at a sample of San Diego residents, comparing talk radio listeners to nonlisteners, and recorded relative levels of information and misinformation. Our results were at once encouraging and somewhat alarming: while talk radio listening is strongly associated with objective information about public affairs, it is also highly correlated with misinformation—the confident holding of information that is objectively false. We found that the distinction depended on whether the information had an ideological dimension. For example, talk radio listeners were more likely than nonlisteners to accurately answer questions involving political information without any kind of ideological element—such as how much of a majority is needed in Congress to override a presidential veto. But those who listened to shows with conservative hosts were also much more likely to inaccurately perceive that the federal budget deficit had grown under the Clinton administration. At the same time, we found that listeners to shows with moderate hosts tended to have the lowest levels of misinformation in the sample.

We do not attribute the high levels of misinformation among conservative-show listeners to lying on the part of conservative talk show hosts, but rather to the tendency on the part of listeners to engage in "inferential reasoning," whereby they receive granules of correct information, combine it with the ideological message they are hearing, and draw inferences about reality. This argument is consistent with the constructionist model of media effects, where media neither powerfully inject the masses with doses of propaganda that the masses unwittingly take in, nor do they only have "minimal effects"—much like Plato's allegory of the cave, message receivers take what they learn, fit it into the image of reality that they have based on personal experience, and draw inferences about reality that may or may not be accurate. Hence the same message can produce different effects in the heads of different people.

So does talk radio enhance the prospects of American deliberative democracy? It appears that it has the potential to enhance civic-mindedness and spur listeners to feel as though they can make a difference in the political process, but outside the experimental laboratory, that community spirit only seems to extend to listeners whose opinions are given voice by those achieving airtime on the shows. Those who are not in agreement with the host or the majority of callers may feel isolated and may choose to silence themselves in the political arena. Furthermore, listeners to ideologically charged shows appear to engage in inferential reasoning that seems to lead, in this case, to greater misinformation and demagoguery, which is antithetical to the goals of deliberative democracy. Hence perhaps a talk radio universe governed by the now-extinct Fairness Doctrine and therefore dedicated to objectivity and the equal expression of different points of view would have great promise as an agent of deliberative democracy. But that is not the universe in which we live—where talk radio is dedicated to maintaining an audience and furthering the interests of the host. Like the other forms of the "new media," maintaining an audience means entertaining the audience and often targeting an audience. Objectivity and equal opportunities for expression do not serve those purposes as well as satire, polemic, and otherwise telling the target audience what it wants to hear.

Media Effects

Given that more and more Americans are receiving information about public affairs from various manifestations of new media, the efforts undertaken in this book to uncover talk radio effects may be emblematic of much media-effects research to come. In some ways, the new media provide great new opportunities for uncovering media effects that have often been difficult to uncover. For example, to a greater extent than the traditional media, the new media often provide messages that are clear, unambiguous, repetitive, simple, convenient, and entertaining. All of these things predict persuasion (Jowett and O'Donnell 1986). However, given the ideological tenor of much of the new media, and its corresponding targeting of audiences, the self-selection inherent on the part of new media consumers poses great analytical challenges. In this book, I have carefully crafted several research designs in such a way as to apply the most stringent tests of causality possible. I have done my best to distinguish between relationships that emerge between talk radio listening and choices that are a function of talk radio, from relationships that are a function of choices. For example, in chapters 3 and 6, I reported results from controlled experiments, isolating the causal variables so that observed differences in attitudes and behavior between groups exposed to the message and groups exposed to the "placebo" cannot be attributed to anything but the message. These provide compelling evidence of effects in the abstract, but cannot be generalized to a larger population because of the artificial testing environment.

Perhaps the most compelling evidence of talk radio persuasion comes in chapter 4, where I apply a multimethod approach to overcoming selection bias in attempting to understand the association between regular Limbaugh listening and political choice. I first performed a content analysis to determine which political issues Limbaugh focused on at the expense of others. I then reasoned that if persuasion is taking place, then the strength of the relationship between listening and conservatism should correspond to how much time Limbaugh devotes to the topic in question. I found an almost perfect linear relationship showing that the more the topic is mentioned on Limbaugh's show, the greater the correlation between listening and conservatism. I then went a step further, retesting this hypothesis by substituting an instrumental variable for Limbaugh listening that was purged of variance that could be attributable to selection bias. Doing this served to dramatize the strength of the relationship between listening and conservatism for topics that Limbaugh emphasizes, such as government social spending; it also magnified the lack of a relationship between Limbaugh listening and conservatism regarding topics that Limbaugh spends little time on, such as abortion. Given that Limbaugh listeners are more likely than ordinary Americans to identify themselves as born-again Christians, the lack of an independent relationship between Limbaugh listening and attitudes toward gay rights and abortion cannot be an indication that only "libertarian" conservatives choose to tune in to Limbaugh's show. Third, in chapters 4 and 6, I took advantage of a panel survey design, which interviewed respondents at repeated intervals to monitor how choices evolved over time in response to Limbaugh listening. I found that people are much more likely to make political choices that mirror Limbaugh admonitions after listening to him for some time. Finally, in chapter 5, I isolated Republicans and looked at primary vote choices in 2000, in order to neutralize any role that partisanship might be playing on listening-choice linkages. I found that Limbaugh listeners were much more likely to prefer Limbaugh's chosen candidate than were nonlisteners, using information that could not have been available to them much in advance of the primary season itself. This finding not only served to add further support to the persuasive potential of new media outlets, opinion leaders, and talk radio in general, but served to update the findings presented in other chapters, providing compelling evidence that talk radio effects were not something limited to an isolated and unique period in the mid-1990s.

As a whole, the findings reported in this book support the notion that new media may not be so new after all, but rather may recall the origins of mass media in the United States, when objective journalism was unheard of, the press was unabashedly partisan, and audiences were split among many targeted alternatives. Future analyses of and normative expositions on the form and function of new media will be able to better unearth the degree to which this sea change in the way Americans receive political information serves to heighten or diminish the quality of democratic discourse, whether the "marketplace of ideas"—where ideas compete for prominence on the basis of merit—is being replaced by an information supermarket characterized by advertising and attention deficit disorder.

A The Limbaugh Message

Frequently Mentioned Issues (Mentioned 300 Days or More)

Government Spending (644 Mentions)

"Government is an enemy that threatens your livelihoods and careers" (16 December 1993).

Media (617)

"I make no bones about the fact that I am a conservative, unlike the dominant media that continues to deny its liberal bias" (4 January 1994). "The dominant media doesn't portray the real truth of today's society" (6 January1994).

Health Care (349)

"The [Clinton health care] plan is about giving the government more power and making you a slave . . . [the plan] promises the simplicity of the tax form, the efficiency of the post office, the bureaucracy of the Department of Agriculture, and the results of rent control" (23 September 1993).

Crime Bill (308)

"The crime bill is just another massive spending program" (14 April 1994). "The crime bill is worthless—it contains meaningless policies and programs, such as midnight basketball games in inner cities" (15 July 1994).

Frequently Mentioned Groups

Environmentalists (520)

The pro-environment movement is "the new home of socialists, driven to attack the instruments of capitalism, which they view as the main threat to the environment" (10 February 1995). "Environmental activists are dunderhead alarmists and prophets of doom—long-haired maggot-infested FM-types" (Jamieson, Capella, and Turow 1996).

Feminists (499)

"Feminists encourage women to abandon traditional values" (3 April 1993). "Feminist leadership is trying to alter basic human nature" (23 April 1993). "Feminism was established as to allow unattractive women easier access to the mainstream of society" (Limbaugh's updated *35 Undeniable Truths of Life* as read on his radio show 18 February 1994; http://rosecity.net/rush/truths.html). "I like the women's movement—from behind" (Jamieson, Capella, and Turow 1996).

Frequently Mentioned Political Figures

President Clinton (722), Hillary (530)

"Clinton's promises are not worth the paper they are written on" (22 September 1993). "[The Clintons] true objective—a bigger and more powerful government which punishes those who achieve and those who are successful. The Clintons want to make people more dependent on government and to equalize everyone by pulling the top down" (30 September 1993).

H. Ross Perot (460)

"Perot is crazy" (9 January 1995). "Perot was the guest host on Larry King Live last night. Evidently CNN decided to move from softball to screwball" (17 January 1995).

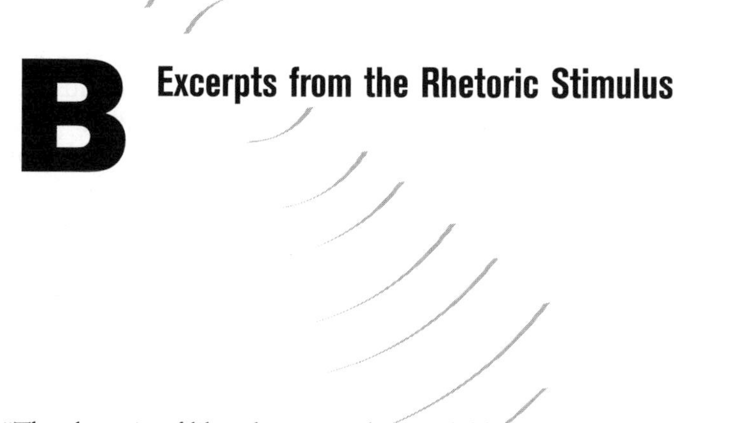

B Excerpts from the Rhetoric Stimulus

"The rhetoric of liberalism is understandably appealing. But when sensible people see over and over again in their lives how these ideas do not work in the real world, they walk away from them."

"World history shows that no nation ever taxed itself into prosperity. There is no way that government redistribution of wealth can create wealth. All over the world, nations are abandoning this idea in favor of market economics. Nations that used to embrace redistribution are now running from it. Why? Because it doesn't work! It inevitably leads to national economic suicide. But Rush, what about the Depression? Didn't the New Deal lead us out of the worst economic slump in history? The answer is an emphatic No! It was World War II that led us out of the Depression, not the New Deal. The Depression began in 1929. During the worst of it, unemployment reached about 25%. In 1939, when the New Deal was fully in force, unemployment was only down to 17%."

"Liberals think that you can't make it without affirmative action by big government to improve your life. These people want to create as much government dependency as possible, not because it will improve our lives, but because it will empower them. For proof of this theory you

---

All quotes taken from *See, I Told You So* (Limbaugh 1993).

need only refer to the fact that whatever liberal legislation these folks impose on society, they inevitably exempt themselves from its rules and effects."

"Polls show that people actually think that liberals care more about homelessness, unemployment, and so on than conservatives do. Not because liberals have actually done anything tangible about these problems, but because their rhetoric is kinder and gentler. That's all liberalism is about—symbolism over substance. It doesn't matter that the policies don't work. 'We care more because we're better people. So keep on voting for us, not those right-wing meanies.' But we have experience with what these 'high-minded' ideals translate into in the real world—huge new bureaucracies, hundreds of billions of dollars worth of new tax increases, price controls, rationing of health care and expansive new regulations. This is the real fallacy of liberalism, that human compassion emanates from government. The only way that liberals have of implementing their involuntary compassion schemes is by micro-management, regulation and the bureaucracy, and Hillary knows it. I am not impeaching the character of all liberals. [Some] have not really thought through the issues of the day. [Some of these people] may be truly well meaning, they may not, however, fully understand the ramifications of the social prescriptions of the left. Liberals respond reflexively to various stimuli. For example, the cure for homelessness equals more taxes. The cure for unemployment equals more taxes. The cure for illiteracy equals more taxes. The answer is always the same—higher taxes. All you have to do to be a good liberal is to say 'yes' to everything, except downsizing government and cutting taxes."

C Excerpts from the Value Heresthetic Stimulus

"Dan's Bake Sale gave me even more confidence in my beliefs and in the values that we share. It was a microcosm of America. It was a group of ordinary Americans who are self-sufficient, and who don't rely on others to make their breaks. This was free enterprise on the march! It all began with a twenty dollar newsletter subscription and look how it blossomed. Plus, Dan was not given a handout. Instead, he had to earn it, and in the process, learn important lessons about free enterprise."

"In a microcosm, it showed the value of initiative and free enterprise, cornerstones of the American dream. . . . Isn't the American Dream about removing the shackles of government and turning loose individuals to create and produce and enjoy the fruits of their own hard work and investment? Wasn't America founded on the principle of people choosing freely to confederate for mutually beneficial services?"

"This was an example of good old American hard work, self-reliance and free enterprise. Real entrepreneurship was demonstrated; real wealth was created. These things should be rewarded."

"People should be empowered to run their own lives and determine their own destiny. They should not be shackled by over-regulation

---

All quotes taken from *See, I Told You So* (Limbaugh 1993).

and the red tape of government intruding into every aspect of their lives."

"I will never stop promoting economics that empower people. I will never give up encouraging people to pursue excellence. I will relentlessly and tirelessly encourage people to be the best they can be. And I want the country to be the best it can be. For that to happen we need strong, self-reliant individuals. We need to reward risk and stop punishing achievement. Let the marketplace work."

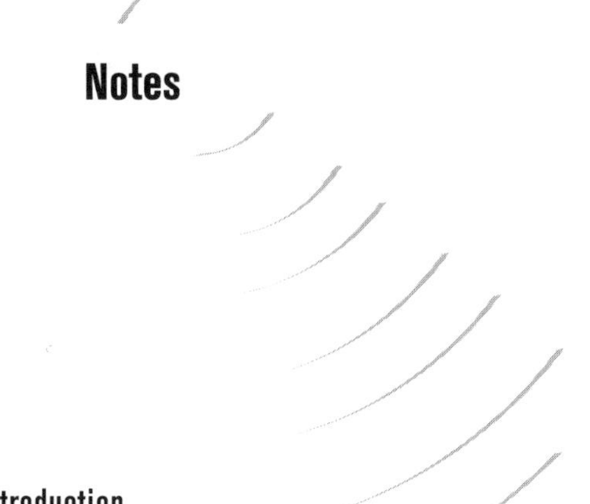

Notes

1 Introduction

1. Of course, many critics would disagree that the traditional mass media are objective. Some argue that, because journalists are predominantly Democratic in their party identification, the media have a liberal bias (e.g., Lichter, Rothman, and Lichter 1986; but cf. Gans 1985). Others argue that because the media are owned by large corporations and depend on advertisers to pay the bills, it often reflects a rightward-leaning pro-business slant (e.g., Bagdikian 1992).

2. The term "heresthetic" was first coined by Riker (1983).

2 Political Talk Radio and Its Most Prominent Practitioner

1. These figures are similar to those reported during the same time period by the American National Election Studies and the Pew Center for the People and the Press.

2. These percentages grew slightly in 1996 and fell slightly in 1997. The slight changes are probably attributable to sampling error. Other surveys by the Pew Center and the Annenberg School of Communication produced comparable results in terms of audience size.

3. Such characteristics would normally be associated with liberalism in economic policy and lower levels of political efficacy and political participation.

4. Rush Limbaugh radio program, July 1996.

5. John Switzer is a private citizen who for four years summarized (in great detail) the specifics of each Limbaugh broadcast and posted the summaries on

the Internet (http://www.math.ukans.edu/atteberr/rsums.html). Switzer stopped creating the summaries during the summer of 1996.

6. This method might be compared to Fan's (1988) filtering of AP wire reports.

3 Toward a Value Heresthetic Model of Political Persuasion

1. Subjects were primarily recruited from Introduction to American Government classes. However, a small number of students were recruited from two upper-division classes: American Political Thought, and Media and Politics.

2. Many thanks go to Robert L. Lineberry, Greg Roof, Ross M. Lence, Dana Ables, and George Antunes for allowing us to recruit students from their classes.

3. The pretest survey is available from the first author. Because of concerns that the survey could cue subjects to the nature of the experiment (Campbell and Stanley 1963), 39 percent of subjects were assigned to the "no pretest" condition. Post-hoc analyses reveal that our concern was unwarranted. The mean differences between the group who received a pretest and the group who did not receive a pretest were statistically and substantively insignificant.

4. The zero-order differences of means on opposition to poor spending according to experimental stimulus were less pronounced due to the disproportionately liberal predispositions of the subjects randomly assigned to receive the value heresthetic stimulus. However, the differences of means were still statistically significant ($p < .05$) in a one-tailed test.

5. This structural equation was created using EQS software.

4 Talk Radio, Public Opinion, and Vote Choice: The "Limbaugh Effect," 1994-96

1. This variable was measured on a five-point scale, coded as follows: 0 < never listens, 1 < listens occasionally, 2 < listens at least once a week, 3 < listens almost every day, 4 < listens every day.

2. In early models, we also tested the interaction between listening to Limbaugh and liking him, but found no systematic effects or marginal utility to the model in terms of changes in variance explained, so we removed this control variable from the model.

3. This opinion was predicted using logistic regression rather than OLS. Thus the coefficients represent the change in the log of the odds of a respondent supporting the crime bill for a one-unit change in the independent variable.

4. Opinions regarding national health insurance, the crime bill, and the women's movement were measured in 1994 rather than 1995, because opinion on these items were not measured by the ANES in 1995. Even though analyz-

ing the relationship between Limbaugh listening in 1995 and opinion toward these topics in 1994 poses a temporal problem, we felt some purchase could be gained by exploring the relationships, because those respondents who reported listening in 1995 were likely listeners in 1994 as well, and those three topics were highly salient and pervasive in Limbaugh's message during 1993–95. Nevertheless, we place less stock in results obtained for relationships that reveal this reversal of temporal order.

5. The opinions reported in this table do not represent all of the topics discussed regularly by Limbaugh for which a strong statistical association between listening and opinion can be observed. The choice to report the associations for these particular topics was not made arbitrarily or selectively. These topics stand out for various idiosyncratic reasons. National health insurance was the most salient political issue (arguably) in 1993 and 1994. The crime bill represents the only opinion queried in the 1994–95 American National Election Studies pertaining to a specific piece of legislation. Limbaugh's comments toward environmentalists and feminists have received significant media attention. As mentioned earlier, the primary focus of Limbaugh's message, in policy terms, has to do with the role of government in the domestic economy. In political terms, Limbaugh's primary target has been President Clinton, by a wide margin, followed by the media. Finally, we make special note of opinion toward Perot because such opinion cannot be predicted by partisanship or ideology, offering an opportunity to control for selection bias. The pattern of relationships found for these variables is representative of the pattern for all of the other opinions we examined. Although we have chosen not to report results for the full equations due to space considerations, the other topics emphasized by Limbaugh for which we predicted opinions, all of which are predicted in a significant way by Limbaugh listening, include Vice President Gore, big business (both raw and relative to labor unions), the government's responsibility to provide jobs, Clinton's ideological position, the trade-off between jobs and the environment, welfare, taxes, the deficit, Republicans, and Democrats.

6. It should be noted that the substance of the two-stage least-squares results matches the results obtained using the raw measure of Limbaugh listening very closely across all variables analyzed, not just the ones presented in this section. The only exception to this rule involves former Senator Bob Dole. While the raw indicator of Limbaugh listening appeared to significantly predict hostility toward Dole, the instrumental indicator reveals no relationship, a more intuitively pleasing finding because Limbaugh's treatment of Dole was ambivalent throughout 1993–95. This discrepancy leads us to conclude that, with respect to Dole, multicollinearity between the independent variables in the earlier models led to a statistical artifact.

6 The Talk Radio Community: Nontraditional Social Networks and Political Participation

1. The size of the sample fell because the first eleven subjects were part of a pilot designed specifically to evaluate the design of the attitude-change experiment.

2. A structural equation analysis performed using EQS modeling software produced substantively identical results to those presented here: exposure to efficacy heresthetic is positively related to both political efficacy and activity, thus revealing both a direct and indirect relationship with participation in the committee debate. Moreover, the model conforms empirically to the data quite well, producing a comparative fit index of .97.

7 Information, Misinformation, and Political Talk Radio

1. Interviewing was conducted in English by students in political behavior classes late spring and early summer, 1997, after training students in the design and purposes of the project and training in telephone interviewing methods. Failure to interview persons due to language occurred for less than 5 percent of potential participants. Up to four callbacks were made, resulting in an overall completion rate of 48 percent, a rate surpassing that of the better survey firms in the area. Distributions for the sample generally corresponded to 1990 U.S. Census data for San Diego, although minorities, less-affluent, and less-educated persons were slightly underrepresented. For the sample, mean age was 41.7 (SD = 16.7) years, mean household income $38,441 (SD = $10,288), and mean education was 14.9 (SD = 2.5) years. About 72.9 percent identified as Caucasian, 6.9 percent African American, 13 percent Latino, and 7.2 percent Asian. About 46 percent were male and 54 percent were female. Politically, 28.2 percent identified as strong or weak Republicans, 11.5 percent leaned Republican, 29.2 percent identified as strong or weak Democrats, 7.7 percent leaned Democratic, and 23.4 percent claimed to be Independent. About 35.9 percent identified as conservative or very conservative, 27.9 percent liberal. On the whole, this sample appears quite similar in terms of demographics to the population of the nation as a whole (American National Election Studies, 1996–97). Where we do see differences, more San Diego residents are Hispanic than we observe in the nation as a whole, and San Diego residents appear to be somewhat less partisan yet more ideological than the national population. However, in terms of the percentages of men versus women, Democrats versus Republicans, liberals versus conservatives, and so on, San Diego looks very much like the nation as a whole, showing no statistically significant differences. This general similarity of the San Diego sample to the nation as a whole lends support to the representativeness of

this sample and to the generalizability of this study.

2. Forty-two percent of the sample reported having "ever listened to talk radio" and 35 percent having listened to talk radio at least once in the last month. Twenty-one percent reported listening to moderate talk radio (NPR, Leykus, or "other") in the last month, while 26 percent reported listening to conservative talk radio (Limbaugh, Liddy, Matalin, or Hedgecock) in the last month. Twenty-one percent of the sample reported having listened to Rush Limbaugh in the last month, and 25 percent reported listening to local conservative Roger Hedgecock. Roughly 10 percent of the sample reported listening at least four times in the last month (or roughly once a week) to either Limbaugh or Hedgecock. By contrast, only 8 percent reported listening to former Watergate plumber G. Gordon Liddy or Bush campaign manager Mary Matalin in the last month, and less than 1 percent reported listening at least four times to these hosts. Roger Hedgecock is a former mayor whose political talk radio program closely resembles that of national host Rush Limbaugh. While the issue content that Hedgecock emphasizes reflects more of a preoccupation with local issues, the style and format of the show follows the Limbaugh model of brash, bombastic entertainment. The similarity of the two commentators is reflected in the nearly identical factor loadings for the shows, (.81 and .78, respectively).

3. Race was measured by asking respondents their race (White, Black, Hispanic, Asian, or "other"). The variable was coded so that "1" indicated that the respondent was white, and "0" indicated that the respondent was "not white." Sex was coded by the interviewer—"1" equals "female." Political interest was measured by asking: "How interested would you say you are in politics and public affairs? Are you very interested, interested, not very interested, or not at all interested?" Partisanship was measured by asking: "Generally speaking, do you usually think of yourself as a strong Republican, a weak Republican, leaning Republican, a strong Democrat, a weak Democrat, leaning Democratic, an independent, or what?" Ideology was measured by asking "Do you usually think of yourself as very conservative, conservative, very liberal, liberal, middle-of-the-road, or don't you think of yourself along these lines?" Income was measured by asking: "About what do you think your total income was last year for yourself and your immediate family before taxes?" Age was measured in years, and education was measured in years completed.

4. We believe that measuring misinformation this way most accurately captures the difference between thinking something is true or thinking something is not true—thus cleanly representing the distinction between information and misinformation. However, we tried a variety of different ways of measuring misinformation, including creating a five-item scale where "2" represented confidence that the statement was true, "0" represented ignorance, and "-2" repre-

sented confidence that the statement was untrue, and summing these responses. Regardless of how we measured misinformation, the substance of our regression results did not change. Therefore, for theoretical reasons just described, we ultimately chose to measure misinformation as a sum of dichotomized responses that items are at least thought to be true, or at least thought to be untrue.

5. Operationally, this study analyzes the relationship between listening to conservative talk radio and believing conservatively charged false statements. However, by no means do we intend to imply that misinformation spreads via conservative talk radio only. Rather, the most popular talk radio programs are extremely conservative, so it seems more relevant to study misinformation from this particular angle. We posit that listening to liberal talk shows would likely result in the spread of misinformation as well, only in the opposite ideological direction.

6. Deleting the political talk activity scale from the model does not significantly alter the substantive relationships between political information and conservative or moderate talk radio listening. Likewise, deleting the exposure scales does not diminish the association between political talk activity and information.

References

Abramson, P. and J. Aldrich. 1982. "The Decline of Electoral Participation in America." *American Political Science Review* 76:502–521.

Adams Research. 1995. "Talk Daily 1995 Talk Radio Survey." *Talk Daily* July/August.

Alesina, A. and H. Rosenthal. 1995. *Partisan Politics, Divided Government, and the Economy*. Cambridge: Cambridge University Press.

Allen, M. 1991. "Meta-Analysis Comparing the Persuasiveness of One-Sided and Two-Sided Messages." *Western Journal of Speech Communication* 55:390–404.

Alvarez, R. M. 1998. *Information and Elections*. Ann Arbor: The University of Michigan Press.

Alwin, D. F., R. L. Cohen, and T. M. Newcomb. 1991. *The Women of Bennington: A Study of Political Orientations over the Life Span*. Madison: University of Wisconsin Press.

American National Election Studies, 1994–97. Ann Arbor, Mich.: University of Michigan, Center for Political Studies (producer and distributor).

Aranson, T. and M. J. Carlsmith. 1963. "Effect of the Severity of Threat on the Devaluation of Forbidden Behavior." *Journal of Abnormal and Social Psychology* 45:584–588.

Areni, C. S. and R. J. Lutz. 1988. "The Role of Argument Quality in the Elaboration Likelihood Model." *Advances in Consumer Research* 15:197–203.

Armstrong, C. B. and A. M. Rubin. 1989. "Talk Radio as Interpersonal Communication." *Journal of Communication* 39:84–94.

Asch, S. E. 1956. "Effects of Group Pressure upon the Modification and Distortion of Judgments." In H. Guetzkow, ed., *Groups, Leadership, and Men*, pp. 177–190. Pittsburgh, Pa.: Carnegie Press.

Avery, R. K. and D. G. Ellis. 1979. "Talk Radio as an Interpersonal Phenomenon." In G. Gumpert and R. Catheart, eds., *Inter/Media: Interpersonal Communication in a Media World,* pp. 108–115. New York: Oxford University Press.

Avery, R. K., D. G. Ellis, and T. W. Glover. 1978. "Patterns of Communication on Talk Radio." *Journal of Broadcasting* 22:5–17.

Bagdikian, B. H. 1992. *The Media Monopoly,* 4th ed. Boston: Beacon Press.

Barber, B. 1984. *Strong Democracy.* Berkeley: University of California Press.

Barker, D. C. 1998a. "Rush to Action: Political Talk Radio and Health Care (un)Reform." *Political Communication* 15:83–97.

———. 1998b. "The Talk Radio Community: Nontraditional Social Networks and Political Participation." *Social Science Quarterly* 79:261–272.

———. 1998c. "Measures of Exposure to Talk Radio." Report to the Board of the American National Elections Studies, 1997 Pilot Study.

———. 1999. "Rushed Decisions: Political Talk Radio and Vote Choice, 1994–96." *Journal of Politics* vol. 61, no. 2:527–539.

Barker, D. C. and C. J. Carman. 2000. "The Spirit of Capitalism? Religious Doctrine, Values, and Economic Attitude Constructs." *Political Behavior* vol. 22, no.1 (March): 1–27.

Barker, D. C. and K. Knight. 2000. "Talk Radio Turns the Tide? Political Talk Radio and Public Opinion." *Public Opinion Quarterly* vol. 64, no. 2 (June):149–170.

Bartels, L. 1988. *Presidential Primaries and the Dynamics of Public Choice.* Princeton: Princeton University Press.

———. 1993. "Messages Received: The Political Impact of Media Exposure." *American Political Science Review* 87:267–285.

———. 1996. "Uninformed Votes: Information Effects in Presidential Elections." *American Journal of Political Science* vol. 1, no. 40:194–230.

Bennett, L. 1980. *Public Opinion in American Politics.* New York: Harcourt, Brace, Jovanovich.

———. 1988. *News: The Politics of Illusion,* 2nd ed. New York: Longman.

Berelson, B., H. Lazarsfeld, and W. N. McPhee. 1954. *Voting.* Chicago: University of Chicago Press.

Berliner, D. C. and B. J. Biddle. 1995. *The Manufactured Crisis: Myths, Fraud, and the Attack on America's Public Schools.* New York: Longman.

Bick, J. 1988. Review of "Talk Radio and the American Dream." *Journal of Broadcasting and Electronic Media* 32:121–122.

Bierig, J. and J. Dimmick. 1979. "The Late Night Radio Talk Show as Interpersonal Communication." *Journalism Quarterly* 56:92–96.

Bolce, L., G. de Maio, and D. Muzzio. 1996. "Dial-In Democracy: Talk Radio and the 1994 Election." *Political Science Quarterly* 111:457–483.

Brinkley, A. 1982. *Voices of Protest: Huey Long, Father Coughlin, and the Great Depression.* New York: Knopf.

Brock, T. C. 1965. "Communication-Recipient Similarity and Decision Change." *Journal of Personality and Social Psychology* 3:296–309.

Brody, R. 1978. "The Puzzle of Political Participation." In A. King, ed., *The New American Political System,* pp. 287–324. Washington, D.C.: American Enterprise Institute for Public Policy Research.

Burke, E. [1790] 1910. *Reflections on the Revolution in France.* London: Everyman's Library.

Cacioppo, J. T., R. E. Petty, and K. J. Morris. 1983. "Effects of Need for Cognition on Message Evaluation, Recall, and Persuasion." *Journal of Personality and Social Psychology* 45:805–818.

Campbell, A., P. E. Converse, W. E. Miller, and D. E. Stokes. 1960. *The American Voter.* New York: John Wiley and Sons.

Campbell, D. T. and J. C. Stanley. 1963. *Experimental and Quasi-experimental Designs for Research.* Boston: Houghton Mifflin.

Capella, J. N., J. Turow, and K.H. Jamieson. 1996. "Call-In Political Talk Radio: Background, Content, Audiences, Portrayal in Mainstream Media." A report from the Annenberg Public Policy Center of the University of Pennsylvania.

Chaffee, S. and J. Scheudler. 1986. "Measurement and Effects of Attention to Media News." *Human Communication Research* 13:76–107.

Chaiken, S. 1986. "Physical Appearance and Social Influence." In C. P. Herman, M. P. Zanna, and E. T. Higgins, eds. *Physical Appearance, Stigma, and Social Behavior: The Ontario Symposium,* vol.5, pp. 3–39. Hillsdale, N.J.: Erlbaum.

Chaiken, S. and A. H. Eagly.1976. "Communication Modality as a Determinant of Message Persuasiveness and Message Comprehensibility." *Journal of Personality and Social Psychology* 34:241–256.

Chaitt, J. 2000. "This Man Is Not a Republican." *The New Republic* (31 January):26–29.

Charters, W. W. 1933. *Motion Pictures and Youth: A Summary.* New York: Macmillan.

Chase, F. 1942. *Sound and Fury: An Informal History of Broadcasting.* New York: Harper and Brothers.

Chase, S. 1956. *Guides to Straight Thinking, with 13 Common Fallacies.* New York: Harper and Row.

Cialdini, R. B., A. Levy, C. P. Herman, L. T. Kozlowski, and R. E. Petty. 1976. "Elastic Shifts of Opinion: Determinants of Direction and Durability." *Journal of Personality and Social Psychology* 34:633–672.

Cialdini, R. B., M. R. Trost, and J. T. Newsom. 1995. "Preference Consistency: The Development of a Valid Measure and the Discovery of Surprising Behavioral Implications." *Journal of Personality and Social Psychology* 69:318–328.

Cobb, M. D. and J. H. Kuklinski. 1997. "Changing Minds: Political Arguments and Political Persuasion." *American Journal of Political Science* 41:88–121.

Combs, J. E. and D. Nimmo. 1993. *The New Propaganda: The Dictatorship of Palaver in Contemporary Politics.* New York: Longman.

Conover, P. J. and S. Feldman. 1984. "Group Identification, Values, and the Nature of Political Beliefs." *American Politics Quarterly* 12:151–177.

Converse, P. E. 1964. "The Nature of Belief Systems in Mass Publics." In D. Apter, ed., *Ideology and Discontent,* pp. 206–261. New York: Free Press.

Conway, M. M. 1985. *Political Participation in the United States.* Washington, D.C.: Congressional Quarterly Press.

Cook, T. D. and D. T. Campbell. 1966. *Quasi-Experimentation: Design and Analysis Issues for Field Settings.* Chicago: Rand McNally College Publishing.

Cooper, H. M. 1979. `statistically Combining Independent Studies: Meta-Analysis of Sex Differences in Conformity Research." *Journal of Personality and Social Psychology* 37:131–146.

Cooper, J. and R. H. Fazio. 1984. "A New Look at Dissonance Theory." In L. Berkowitz, ed., *Advances in Experimental Psychology* 17, pp. 229–266. New York: Academic Press.

Crigler, A., ed. 1996. *The Psychology of Political Communication.* Ann Arbor: University of Michigan Press.

Crittenden, J. 1971. "Democratic Functions of the Open Mike Radio Forum." *Public Opinion Quarterly* 35:200–210.

Dahl, R. A. 1961. *Who Governs? Democracy and Power in an American City.* New Haven, Conn.: Yale University Press.

———. 1989. *Democracy and its Critics.* New Haven, Conn.: Yale University Press.

Dalton, R. J., P. A. Beck, and R. Huckfeldt. 1998. "Partisan Cues and the Media: Information Flows in the 1992 Presidential Election." *American Political Science Review* 92:111–126.

Davis, K. C. 1992. *Don"t Know Much about Geography: Everything You Need to Know about the World but Never Learned.* New York: W. Morrow.

Davis, R. and D. Owen .1998. *New Media and American Politics.* New York: Oxford University Press.

DeBono, K. G. 1987. "Investigating the Social-Adjustive and Value-Expressive Functions of Attitudes: Implications for Persuasion Processes." *Journal of Personality and Social Psychology* 52:279–287.

De Fleur, M. L. and S. Ball-Rokeach. 1988. *Theories of Mass Communication.* New York: Longman.

Delli Carpini, M. X. and S. Keeter. 1996. *What Americans Know about Politics and Why It Matters.* New Haven, Conn.: Yale University Press.

Devroy, A. and K. Merida.1994. "Angry President Assails Radio Talk Shows." *The Washington Post* (25 June):A1.

Dorman, W. A. and S. Livingston. 1994. "News and Historical Content: The Establishing Phase of the Persian Gulf Policy Debate." In W. L. Bennett, ed., *Taken by Storm,* pp.63–81. Chicago: University of Chicago Press.

Dreier, P. and W. J. Middleton. 1994. "How Talk Radio Helped GOP's Resurgence." *Chicago Tribune* (21 December):129

Eagly, A. H. and S. Chaiken. 1993. *The Psychology of Attitudes.* Forth Worth, Tex.: Harcourt, Brace, Jovanovich.

Easton, D. 1965. *A Framework for Political Analysis.* Englewood Cliffs, N.J.: Prentice-Hall.

Edsall, T. B. and T. M. Neal. 2000. "Bush, Allies Hit McCain's Conservative Credentials." *Washington Post* (15 February):A1.

Entman, R. 1993. "Framing U.S. Coverage of International News: Contrasts in Narratives of the KAL and Iran Air Incidents." *Journal of Communication* 41:6–27.

Fan, D. P. 1988. *Predictions of Public Opinion from the Mass Media.* New York: Greenwood Press.

Feldman, S. 1983. "Economic Individualism in American Public Opinion." *American Politics Quarterly* 11:3–29.

———. 1988. `structure and Consistency in Public Opinion: The Role of Core Beliefs and Values." *American Journal of Political Science* 32:416–440.

Ferguson, R. F. 1991. "Paying for Public Education: New Evidence on How and Why Money Matters." *Harvard Journal on Legislation* 28:465–498.

Festinger, L. 1957. *A Theory of Cognitive Dissonance.* Evanston, Ill.: Row, Peterson.

Festinger, L. and N. Macoby. 1964. "On Resistance to Persuasive Communication." *Journal of Abnormal and Social Psychology* 68:359–366.

Finifter, A. W. 1974. "The Friendship Group as a Protective Environment for Political Deviants." *American Political Science Review* 68:607–625.

Fishbein, M. and I. Ajzen. 1981. "Acceptance, Yielding, and Impact: Cognitive processes in Persuasion." In R. E. Petty, T. M. Ostrom, and T. C. Brock, eds., *Cognitive Responses in Persuasion,* pp. 339–359. Hillsdale, N.J.: Erlbaum.

Fishkin, J. S. 1995. *The Voice of the People.* New Haven, Conn.: Yale University Press.

Freedman, J. and D. Sears. 1965. "Warning, Distraction, and Resistance to Influence." *Journal of Personality and Social Psychology* 1:262–266.

Gabler, N. 1995. "A Multitude of Meanness." *Los Angeles Times* (1 January):M1.

Gamson, W. A. and A. Modigliani. 1989. "Media Discourse and Public Opinion on Nuclear Power: A Constructionist Approach." *American Journal of Sociology* 95:1–37.

Gans, H. J. 1985. "Are U.S. Journalists Dangerously Liberal?" *Columbia Journalism Review* 24:29–33.

Gianos, C. and C. R. Hofstetter. 1995. "Participation, Involvement, and Political Entertainment: A Community Study of Political Talk Radio." Paper presented at the Annual Meeting of the Southern Political Science Association, 1–4 November, Tampa, Fla.

Gitlin, T. 1980. *The Whole World is Watching: Mass Media in the Making and Unmaking of the New Left.* Berkeley: University of California Press.

Goldberg, P. 1968. "Are Women Prejudiced Against Men?" *Transaction* 5:28–30.

Grunig, J. E. 1989. "Publics, Audiences, and Market Segments: Segmentation Principles for Campaigns." In C. T. Salmon, ed., *Information Campaigns: Communications Research,* pp. 199–228. Beverly Hills, Calif.: Sage Publications.

Gurin, P., G. Gurin, R. Lao, and M. Beattie. 1969. "Internal-External Control in the Motivational Dynamics of Negro Youth." *Journal of Social Issues* 25:29–58.

Hallin, D. C. 1995. "The "Uncensored War.' "' In J. P. Vermeer, ed., *In "Media" Res: Reading in Mass Media and American Politics,* pp. 11–21. New York: McGraw-Hill.

Hammond, T. H. and B. D. Hume. 1993. " "What This Campaign Is All about Is . . .": A Rational Choice Alternative to the Downsian Spatial Model of Elections." In B. Grofman, ed., *Information, Participation, and Choice: An Economic Theory of Democracy in Perspective.* Ann Arbor: University of Michigan Press.

Hass, R. G. 1981. "Effects of Source Characteristics on Cognitive Responses and Persuasion." In R. E. Petty, T. M. Ostrom, and T. C. Brock, eds., *Cognitive Responses in Persuasion,* pp. 141–172. Hillsdale, N.J.: Erlbaum.

Hass, R. G. and K. Grady. 1975. "Temporal Delay, Type of Fore-Warning, and Resistance to Influence." *Journal of Experimental Psychology* 11:459–469.

Haugvelt, C. P. and D. T. Wegener. 1994. "Message Order Effect in Persuasion: An Attitude Strength Perspective." *Journal of Consumer Research* 21:205–218.

Herbst, S. 1995. "Electronic Public Space: Talk Shows in Theoretical Perspective." *Political Communication* 12:263–274.

Hetherington, M. 1996. "The Media's Role in Forming Voters" National Economic Evaluations in 1992." *American Journal of Political Science* 40:372–395.

Hibbs, G. L. 2000. "McCain Campaign—R.I.P." *NewsSynthesis.com*. On-line: <www.newssynthesis.com/columns/030200.htm>.

Hochschild, J. L. 1981. *What's Fair? American Beliefs About Distributive Justice*. Cambridge: Harvard University Press.

———. 1995. *Facing Up to the American Dream: Race, Class, and the Soul of the Nation*. Princeton: Princeton University Press.

Hofstetter, C. R. 1996. 'situational Involvement and Political Mobilization: Political Talk Radio and Political Action." Paper presented at the annual Meeting of the Midwest Political Science Association, 18–20 April, Chicago.

———. 1998. "Political Talk Radio, Situational Involvement, and Political Mobilization." *Social Science Quarterly* 79:273–286.

Hofstetter, C. R. and C. L. Gianos. 1997. Political Talk Radio: Actions Speak Louder than Words." *Journal of Broadcasting and Electronic Media* 41: 501–515

Hofstetter, C. R., M. Donovan, M. Klauber, A. Cole, C. Huie, and T. Yuasa. 1994. "Political Talk Radio: A Stereotype Reconsidered." *Political Research Quarterly* 47:467–479.

Hofstetter, C. R., J. T. Smith, G. M. Zari, and C. H. Hofstetter. 1997. "The Political Talk Radio Experience: A Community Study." Paper presented at the 1997 Midwest Political Science Association, 10–12 April, Chicago.

Hollander, B. L. 1995. "Talk Radio, the New News, and the Old News." Paper prepared for the Southern Political Science Association, Tampa, Fla.

———. 1996. "Talk Radio: Predictors of Use and Effects on Attitudes about Government." *Journalism and Mass Communication Quarterly* 73:102–113.

Hovland, C. I., O. J. Harvey, and M. Sherif. 1954. "Assimilation and Contrast in Communication and Attitude Change." *Journal of Abnormal and Social Psychology* 55:242–252.

Hovland, C. I., I. L. Janis, and H. H. Kelley. 1953. *Communication and Persuasability*. New Haven, Conn.: Yale University Press.

Hovland, C. I. and M. J. Rosenberg. 1960. *Attitude Organization and Change*. New Haven, Conn.: Yale University Press.

Hovland, C. I. and W. Weiss. 1951. "The Influence of Source Credibility on Communication Effectiveness." *Public Opinion Quarterly* 15:635–650.

Hoyle, R. H., ed. 1995. *Structural Equation Modeling: Concepts, Issues, and Applications*. Thousand Oaks, Calif.: Sage Publications

Huckfeldt, R. and J. Sprague. 1987. "Networks in Context: The Social Flow of Political Information." *American Political Science Review* 81:1197–1216.

———. 1993. "Citizens, Contexts, and Politics." In A. W. Finifter, ed., *Political Science: The State of the Discipline II*. Washington, D.C.: APSA.

Hurwitz, J. and M. A. Peffley 1987. "How are Foreign Policy Attitudes Structured? A Hierarchical Model." *American Political Science Review* 81:1099–1120.

Hurwitz, J., M. A. Peffley, and M. A. Seligson. 1993. "Foreign Belief Systems in Comparative Perspective: The United States and Costa Rica." *International Studies Quarterly* 37:245–270.

Institute for Propaganda Analysis. 1939. *The Fine Art of Propaganda: A Study of Father Coughlin's Speeches,* A. McClung Lee and E. Briant Lee, eds. New York: Harcourt Brace and Company.

Iyengar, S. 1991. *Is Anyone Responsible?* Chicago: University of Chicago Press.

Iyengar, S., and D. R. Kinder. 1987. *News That Matters: Agenda-Setting and Priming in a Television Age.* Chicago: University of Chicago Press.

Jamieson, K. H., J. N. Capella, and J. Turow. 1996. "Limbaugh: The Fusion of Party Leader and Partisan Mass Medium." Paper delivered at the 1996 Annual Meeting of the American Political Science Association, August, San Francisco.

Jamieson, D. W. and M. P. Zanna. 1989. "Need for Structure in Attitude Formation and Expression." In A. R. Pratkanis, S. J. Breckler, and A. G. Greenwald, eds., *Attitude Structure and Function,* pp. 383–406. Hillsdale, N.J.: Erlbaum.

Jones, D. A. 1997. "Political Talk Radio as a Forum for Intra-Party Debate." Paper delivered at the 1997 Southern Political Science Association Annual Meeting, 5–8 November, Norfolk, Va.

Jowett, G. S. and V. O"Donnell, 1986. *Propaganda and Persuasion.* London: Sage Publishing.

Katz, E. and P. F. Lazarsfeld. 1964. *Personal Influence: The Part Played by People in the Flow of Mass Communications.* New York: Free Press.

Katz, I. and R. G. Hass. 1988. "Racial Ambivalence and American Value Conflict: Correlational and Priming Studies of Dual Cognitive Structures." *Journal of Personality and Social Psychology* 55:893–905.

Kernell, S. 1986. *Going Public: New Strategies of Presidential Leadership.* Washington, D.C.: CQ Press.

Kinder, D. R. 1983. "Diversity and Complexity in American Public Opinion." In A. Finifter, ed., *Political Science: The State of the Discipline,* pp. 391–401. Washington, D.C.: American Political Science Association.

———. 1998. "Opinion and Action in the Realm of Politics." In D. T. Gilbert, S. T. Fiske, and G. Lindzey, eds., *The Handbook of Social Psychology,* 4th ed., pp.778–867. New York: McGraw-Hill.

King, G. 1989. *Unifying Political Methodology: The Likelihood Theory of Statistical Inference.* Cambridge: Cambridge University Press.

Klapper, J. 1960. *The Effects of Mass Communication.* New York: Free Press.

Kluckhohn, C. 1951. "Values and Value-Orientations in the Theory of Action." In T. Parsons and E. Shils, eds., *Toward a General Theory of Action,* pp. 388–433. Cambridge: Harvard University Press.

Kluegel, J. R. and E. R. Smith. 1986. *Beliefs about Inequality: Americans' Views about What Is and What Ought to Be.* New York: A. de Gruyter.

Knight, K. and D. C. Barker. 1996. "Talk Radio Turns the Tide? The Limbaugh Effect, 1993–95." Paper presented at the annual meeting of the American Political Science Association, San Francisco.

Krosnick, J. A. and D. R. Kinder. 1990. "Altering the Foundations of Popular Support for the President through Priming." *American Political Science Review* 84:497–512.

Kuklinski, J. H., P. J. Quirk, D. Schweider, and R. F. Rich. 1997. "Misinformation and the Currency of Citizenship." Paper prepared for the Annual Meeting of the Midwest Political Science Association, 10–12 April, Chicago.

Kurtz, H. 1996. *Hot Air: All Talk All the Time.* New York: Times Books.

Lane, R. E. 1962. *Political Ideology.* New York: Free Press.

———. 1973. "Patterns of Political Belief." In J. Knutson, ed., *Handbook of Social Psychology,* pp. 83–116. San Francisco: Jossey-Bass.

Lasswell, H. D. 1958. *Politics: Who Gets What, When, How?* 2nd ed. New York: Meridian Books.

Lasswell, H. D., R. D. Casey, and B. L. Smith. 1935. *Propaganda and Promotional Activities.* Minneapolis: University of Minnesota Press.

Leege, D. C. and L. A. Kellstedt. 1993. *Rediscovering the Religious Factor in American Politics.* Armonk, N.Y.: M. E. Sharpe.

Leighley, J. E. 1990. 'social Interaction and Contextual Influences on Political Participation." *American Politics Quarterly* 18:459–475.

Levin, M. B. 1987. *Talk Radio and The American Dream.* Lexington, Mass.: Lexington Books.

Lichter, S. R., S. Rothman, and L. S. Lichter. 1986. *The Media Elite.* Bethesda, Md.: Adler and Adler.

Limbaugh, R. H. 1992.*The Way Things Ought to Be.* New York: Pocket Books.

———. 1993. *See, I Told You So.* New York: Pocket Books.

———. 1994. "Why Liberals Fear Me." *Policy Review* 70:4–10.

———. 1998. "How to Stay Prosperous and Free in the Twenty-first Century." Special mailing to subscribers of *The Limbaugh Letter.*

Lippmann, W. 1922. *Public Opinion.* New York: Macmillan.

———. 1925. *The Phantom Public.* New York: Harcourt, Brace.

Lochner, L., ed. 1970. *The Goebbels Diaries.* Westport, Conn.: Greenwood.

Lodge, M. and K. M. McGraw. 1995. *Political Judgment: Structure and Process.* Ann Arbor: University of Michigan Press.

Lull, J. 1990. *Inside Family Viewing.* London: Routledge.

Lupia, A. and M. D. McCubbins. 1998. *Democratic Dilemma: Can Citizens Learn What They Need to Know?* New York: Cambridge University Press.

McClosky, H. and J. Zaller. 1984. *The American Ethos: Public Attitudes Toward Capitalism and Democracy.* Cambridge, Mass.: Harvard University Press.

McGuire, W. J. 1968. "Personality and Attitude Change: An Information-Processing Theory." In A. G. Greenwald, T. C. Brock, and T. M. Ostrom, eds. *Psychological Foundations of Attitudes,* pp. 171–196. New York: Academic.

———. 1969. "The Nature of Attitudes and Attitude Change." In G. Lindzey and E. Aranson, eds., *Handbook of Social Psychology,* 2nd ed., vol. 3, pp. 136–314. Reading, Mass.: Addison-Wesley.

———. 1985. "Attitudes and Attitude Change." In G. Lindzey and E. Aranson, eds., *Handbook of Social Psychology,* vol. 2, pp. 233–346. New York: Random House.

Meyerowitz, B. E. and S. Chaiken. 1987. "The Effect of Message Framing on Breast Self-Examination Attitudes, Intentions, and Behavior." *Journal of Personality and Social Psychology* 52:500–510.

Milburn, M. A. 1991. *Persuasion and Politics: The Social Psychology of Public Opinion.* Pacific Grove, Calif.: Brooks/Cole.

Miller, C. R. [1937] 1967. "How to Detect Propaganda." In N. A. Ford, ed., *Language in Uniform.* New York: Odyssey

Miller, N., G. Maruyama, R. Beaber, and K. Valone. 1976. 'speed of Speech and Persuasion." *Journal of Personality and Social Psychology* 34:615–625.

Morris, D. 1998. *Behind the Oval Office: Getting Reelected Against All Odds.* Los Angeles: Renaissance Books.

Mutz, D. C. 1997. "Mechanisms of Momentum: Does Thinking Make It So?" *Journal of Politics* 59:104–125.

Nelson, T. E., R. A. Clawson, and Z. M. Oxley. 1997. "Media Framing of a Civil Liberties Conflict and Its Effect on Tolerance." *American Political Science Review* 91:567–583.

Nelson, T. E. and D. R. Kinder. 1996. "Issue Frames and Group Centrism in American Public Opinion." *Journal of Politics* 58:1055–1078.

Neuman, W. R., M. R. Just, and A. N. Crigler. 1992. *Common Knowledge: News and the Construction of Political Meaning.* Chicago: University of Chicago Press.

Newhagen, J. E. 1994. "Media Use and Political Efficacy: The Suburbanization of Race and Class." *Journal of the American Society for Information Science* 45:386–394.

———. 1996. "Interactivity as a Factor in the Assessment of Political Call-In Programs." Presented at the Annual Meeting of the International Society of Political Psychology, 29 June–3 July, Vancouver, British Columbia, Canada.

Nie, N. H., S. Verba, and J. R. Petrocik 1976. *The Changing American Voter*. Cambridge: Harvard University Press.

Noelle-Neumann, E. 1984. *The Spiral of Silence: Public Opinion—My Social Skin*. Chicago: University of Chicago Press.

Norusis, M. J. 1994. *SPSS Advanced Statistics 6.1*. Chicago: SPSS.

Olson, J. M. and M. P. Zanna. 1993. "Attitude and Attitude Change." *Annual Review of Psychology* 44:117–154.

Owen, D. 1995. "Talk Radio." In J. P. Vermeer, ed., *In "Media" Res: Readings in Mass Media and American Politics,* pp. 60–67. New York: McGraw-Hill.

———. 1997. "Talk Radio and Evaluations of President Clinton." *Political Communication* 14:333–353.

Page, B. I. 1996. *Who Deliberates? Mass Media and Modern Democracy*. Chicago: University of Chicago Press.

Page, B. I. and R. Y. Shapiro. 1983. "Effects of Public Opinion on Public Policy." *The American Political Science Review* 77:175–190.

———. 1992. *The Rational Public*. Chicago: University of Chicago Press.

Page, B. I., R. Y. Shapiro, and G. R. Dempsey. 1987. "What Moves Public Opinion?" *American Political Science Review* 81:23–43.

Paletz, D. L. 1998. *The Media in American Politics: Contents and Consequences*. New York: Longman.

Pan, Z. and G. M. Kosicki. 1993. "Framing Analysis: An Approach to News Discourse." *Political Communication* 10:55–75.

Patterson, T. E. 1993. *Out of Order*. New York: Knopf.

Peffley, M. A., and J. Hurwitz, 1985. "A Hierarchical Model of Attitude Constraint." *American Journal of Political Science* 29:871–890.

Petty, R. E. and T. C. Brock. 1976. "Effects of Responding or not Responding to Hecklers on Audience Agreement with a Speaker." *Journal of Applied Social Psychology* 6:1–17.

Petty, R. E. and J. T. Cacioppo. 1981. *Attitudes and Persuasion: Classic and Contemporary Approaches*. Dubuque, Iowa: W. C. Brown.

———. 1984. "The Effects of Involvement on Responses to Argument Quantity and Quality: Central and Peripheral Routes to Persuasion." *Journal of Personality and Social Psychology* 46:69–81.

Petty, R. E. and J. A. Krosnick, eds. 1995. *Attitude Strength: Antecedents and Consequences*. Mahwah, N.J.: Erlbaum.

Petty, R. E., D. W. Schumann, S. A. Richman, and A. J. Strathman. 1993. "Positive Mood and Persuasion: Different Roles for Affect under High- and Low-Elaboration Conditions." *Journal of Personality and Social Psychology* 64:5–20.

Petty, R. E. and D. T. Wegener. 1998. "Attitude Change: Multiple Roles for Persuasion Variables." In D. T. Gilbert, S. T. Fiske, and G. Lindzey, eds., *The Handbook of Social Psychology,* 4th ed., pp.320–390. New York: McGraw-Hill.

Popkin, S. 1991. *The Reasoning Voter.* Chicago: University of Chicago Press.

Price, V. and J. Zaller. 1993. "Who Gets News? Alternative Measures of News reception and Their Implications for Research." *Public Opinion Quarterly* 57:133–164.

Putnam, R. 1993. *Making Democracy Work: Civic Traditions in Modern Italy.* Princeton: Princeton University Press.

Ratner, E. 1995. "Talk Radio Responds: Our 'Back Fence.'" *Los Angeles Times* (25 April):B7.

Rhodes, N. and W. Wood. 1992. "Self-Esteem and Intelligence Affect Influenceability: The Mediating Role of Message Reception." *Psychological Bulletin* 111:156–171.

Riker, W. H. 1983. "Political Theory and the Art of Heresthetics." In A. Finifter, ed, *Political Science: The State of the Discipline.* Washington, D. C.: American Political Science Association.

———. 1986. *The Art of Political Manipulation.* New Haven, Conn.: Yale University Press.

———. 1990. "Heresthetic and Rhetoric in the Spatial Model." In J. M. Enelow and M. J. Hinich, eds., *Advances in the Spatial Theory of Voting.* Cambridge: Cambridge University Press.

———. 1996. *Strategy of Rhetoric: Campaigning for the American Constitution.* New Haven, Conn.: Yale University Press.

Rokeach, M. 1973. *The Nature of Human Values.* New York: Free Press.

Root, J. 1995. "Is a Picture Worth a Thousand Words? A Quantitative Methodology Study of Political Cartoons." Ph.D. diss., University of Houston Department of Political Science.

Schattschneider, E. E. 1960. *The Semi-Sovereign People: A Realist's View of Democracy in America.* New York: Holt, Rinehart, and Winston.

Schoemaker, P. J., C. Schooler, and W. A Danielson. 1989. "Involvement with the Media." *Communication Research* 16:78–90.

Sears, D. O. 1986. "College Sophomores in the Laboratory: Influences of a Narrow Database on Social Psychology's View of Human Nature." *Journal of Personality and Social Psychology* 51:515–530.

Sears, D. O. and J. L. Freedman. 1967. 'selective Exposure to Information: A Critical Review." *Public Opinion Quarterly* 31:194–214.

Segal, J. and H. J. Spaeth. 1993. *The Supreme Court and the Attitudinal Model.* New York: Cambridge University Press.

Shaw, D. L. and M. E. McCombs. 1977. *The Emergence of American Political Issues: The Agenda Setting Function of the Press.* St. Paul: West Publishing Company.

Sherif, M. and C. I. Hovland. 1961. *Social Judgment: Assimilation and Contrast Effects in Communication and Attitude Change.* New Haven, Conn.: Yale University Press.

Shneidman, E. S. 1969. "Logical Content Analysis: An Explication of Styles of Concludifying." In G. Gerbner, O. R. Holsti, K. Krippendorff, W. J. Paisley, and P. P. Stone, eds., *The Analysis of Communication Content: Developments in Scientific Theories and Computer Techniques,* pp. 261–279. New York: John Wiley and Sons.

Slusher, M. P. and C. A. Anderson. 1996. "Using Causal Persuasive Arguments to Change Beliefs and Teach New Information: The Mediating Role of Explanation Availability and Evaluation Bias in the Acceptance of Knowledge." *Journal of Educational Psychology* 88:110–122.

Smith, E. R. A. N. 1989. *The Unchanging American Voter.* Berkeley: University of California Press.

Sniderman, P. M. 1993. "The New Look in Public Opinion Research." In A. Finifter, ed., *Political Science: The State of the Discipline II.* Washington, D.C.: American Political Science Association.

Sniderman, P. M. and R. A. Brody. 1977. "Coping: The Ethic of Self-Reliance." *American Journal of Political Science* 21:501–522.

Sniderman, P. M., B. K. Wolfinger, D. C. Mutz, and L. E. Wiley. 1991. "Values under Pressure: AIDS and Civil Liberties." In P. M. Sniderman, R. A. Brody, and P. E. Tetlock, eds., *Reasoning and Choice: Explorations in Political Psychology,* pp. 31–57. Cambridge: Cambridge University Press.

Snyder, M. and M. Rothbart. 1971. "Communicator Attractiveness and Opinion Change." *Canadian Journal of Behavioral Science* 3:377–387.

Steele. C. M. 1988. The Psychology of Self-Affirmation: Sustaining the Integrity of the Self." In L. Berkowitz, ed., *Advances in Experimental Social Psychology,* vol. 21, pp. 261–302. New York: Academic Press.

Stimson, J. A., M. B. MacKuen, and R. S. Erikson. 1995. "Dynamic Representation." *American Political Science Review* 89:543–556.

Stoker, L. 1992. "Interests and Ethics in Politics." *American Political Science Review* 86:369–380.

Stoker, L. and M. K. Jennings. 1995. "Life Cycle Transitions and Political Participation: The Case of Marriage." *American Political Science Review* 89:421–433.

Stone, D. 1988. *Policy Paradox.* New York: W. W. Norton.

Surlin, S. H. 1986. "Uses of Jamaican Talk Radio." *Journal of Broadcasting and Electronic Media* 30:459–466.

Tetlock, P. E. 1986. "A Value Pluralism Model of Ideological Reasoning." *Journal of Personality and Social Psychology* 50:819–827.

Texiera, R. 1987. *Why Americans Don"t Vote.* New York: Greenwood Press.

Tocqueville, A. [1848] 1945. *Democracy in America.* New York: Knopf.

Tourangeau, R. and K. A. Rasinski. 1988. "Cognitive Processes Underlying Context Effects in Attitude Measurement." *Psychological Bulletin* 103:299–314.

Traugott, M., A. Berinsky, K. Cramer, M. Howard, R. Mayer, H. P. Schuckman, D. Tewksbury, and M. Young. 1996. "The Impact of Talk Radio on its Audience." Paper presented at the Annual Meeting of the Midwest Political Science Association, 18–21 April, Chicago.

Tull, C. J. 1965. *Father Coughlin and the New Deal.* Syracuse: Syracuse University Press.

Verba, S. and N. Nie. 1972. *Participation in America.* New York: Harper and Row.

Weaver, D. H. 1996. "What Voters Learn about Media." *Annals of the American Academy of Political and Social Science* 546:34–47.

Weber, B. 1992. "A Loud Angry World on the Dial." *The New York Times* (7 June):31.

Weisberg, H., J. Krosnick, and B. Bowen. 1989. *An Introduction to Survey Research and Data Analysis,* 2nd ed. Boston: Scott, Foresman, and Company.

Weschler, L. 1983. "A State of War!" *The New Yorker* (11 April):45–102.

Wolfinger, R. and S. Rosenstone. 1980. *Who Votes?* New Haven, Conn.: Yale University Press.

Wood, W., N. Rhodes, and M. Biek. 1995. "Working Knowledge and Attitude Strength: An Information Processing Analysis." In R. E. Petty and J. A. Krosnick, eds., *Attitude Strength: Antecedents and Consequences,* pp. 283–313. Mahwah, N.J.: Erlbaum.

Zaller, J. R. 1992. *The Nature and Origin of Mass Opinion.* New York: Cambridge University Press.

———. 1996. "The Myth of Massive Media Impact Revisited: New Support for a Discredited Idea." In D. C. Mutz, P. M. Sniderman, and R. A. Brody, eds., *Political Persuasion and Attitude Change*. Ann Arbor: University of Michigan Press.

———. 2001a. "The Mass Media, Party Elites, and the Evolution of the Presidential Nomination System, 1972–2000." Paper presented at the Annual Meeting of the Midwest Political Science Association, 18–22 April, Chicago.

———. 2001b. "Parties Are Back." Paper presented at the Annual Meeting of the Midwest Political Science Association, 24 April, Chicago.

Zaller, J. R. and S. Feldman. 1992. "A Simple Theory of the Survey Response." *American Journal of Political Science*. Vol. 36 (August): 579–616.

Zerbinos, E. 1993. "Talk Radio: Motivation or Titillation?" Paper Presented at the Annual Convention of the Association for Education in Journalism and Mass Communication.

Index

Power, Conflict, and Democracy: American Politics Into the Twenty-first Century

John G. Geer, *From Tea Leaves to Opinion Polls: A Theory of Democratic Leadership*

Kim Fridkin Kahn, *The Political Consequences of Being a Woman: How Stereotypes Influence the Conduct and Consequences of Political Campaigns*

Kelly D. Patterson, *Political Parties and the Maintenance of Liberal Democracy*

Dona Cooper Hamilton and Charles V. Hamilton, *The Dual Agenda: Race and Social Welfare Policies of Civil Rights Organizations*

Hanes Walton Jr., *African-American Power and Politics: The Political Context Variable*

Amy Fried, *Muffled Echoes: Oliver North and the Politics of Public Opinion*

Russell D. Riley, *The Presidency and the Politics of Racial Inequality: Nation-Keeping from 1831 to 1965*

Robert W. Bailey, *Gay Politics, Urban Politics: Identity and Economics in the Urban Setting*

Ronald T. Libby, *ECO-WARS: Political Campaigns and Social Movements*

Donald Grier Stephenson Jr., *Campaigns and the Court: The U.S. Supreme Court in Presidential Elections*

Kenneth Dautrich and Thomas H. Hartley, *How the News Media Fail American Voters: Causes, Consequences, and Remedies*

Douglas C. Foyle, *Counting the Public In: Presidents, Public Opinion, and Foreign Policy*

Ronald G. Shaiko, *Voices and Echoes for the Environment: Public Interest Representation in the 1990s and Beyond*

Demetrios James Caraley, editor, *The New American Interventionism: Lessons from Successes and Failures—Essays from* Political Science Quarterly

Ellen D. B. Riggle and Barry L. Tadlock, editors, *Gays and Lesbians in the Democratic Process: Public Policy, Public Opinion, and Political Representation*

Hanes Walton Jr., *Reelection: William Jefferson Clinton as a Native-Son Presidential Candidate*

Robert Y. Shapiro, Martha Joynt Kumar, Lawrence R. Jacobs, Editors, *Presidential Power: Forging the Presidency for the Twenty-First Century*

Marissa Martino Golden, *What Motivates Bureaucrats? Politics and Administration During the Reagan Years*

Geoffrey Layman, *The Great Divide: Religious and Cultural Conflict in American Party Politics*

Kerry L. Haynie, *African American Legislators in the American States*

Sally S. Cohen, *Championing Child Care*

Lisa J. Disch, *The Tyranny of the Two-Party System*

David C. Barker, *Rushed to Judgment?: Talk Radio, Persuasion, and American Political Behavior*